Geometry Labs

Henri Picciotto

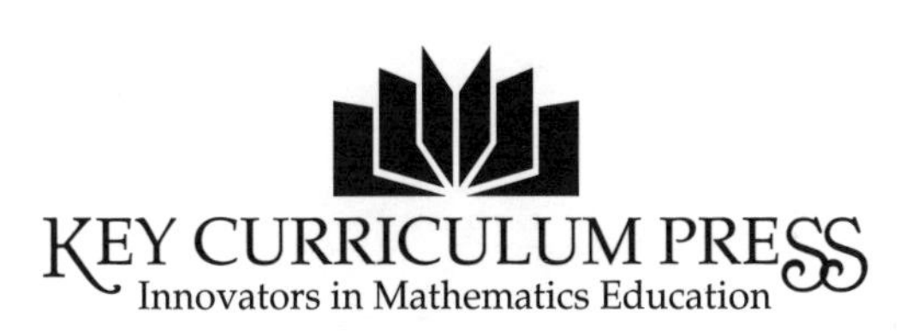

Project Editor: Dan Bennett
Editorial Assistant: James A. Browne
Production Editor: Jason Luz
Copy Editor: Paul Green
Production and Manufacturing Manager: Diana Jean Parks
Production Coordinator: Ann Rothenbuhler
Text Designer: Kirk Mills
Compositor: Ann Rothenbuhler
Cover Designer and Illustrator: Diane Varner
Technical Artist: Kirk Mills
Prepress and Printer: Versa Press, Inc.

Executive Editor: John Bergez
Publisher: Steven Rasmussen

Key Curriculum Press
1150 65th Street
Emeryville, CA 94608
510-595-7000
editorial@keypress.com
http://www.keypress.com

Printed in the United States of America 10 9 8 7 6 5 4 3 2 1 03 02 01 00 99 ISBN 1-55953-361-7

Acknowledgments

Many of these activities were developed at the Urban School of San Francisco. Special thanks to my students and colleagues there, particularly Richard Lautze and Kim Seashore.

These math teachers and professors helped develop my love for geometry, or offered insights which have no doubt found their way into this book: Cal Crabill, G. D. Chakerian, Lew Douglas, Phil Mallinson, Sherman Stein, and Joel Teller.

Finally, I'd like to thank the authors of the Elementary Science Study at the Educational Development Center, whose math units *Tangrams* and *Pattern Blocks* awakened in me an interest in doing math with manipulatives when I was a beginning teacher, many years ago.

Henri Picciotto

Contents

Introduction

About This Book

This book is a collection of activities in secondary-school geometry. Most of the activities are hands-on and involve concrete materials. Many of them have enough depth to provide excellent opportunities for discussion and reflection about subtle and important ideas. Like others of my books, this one is not geared to a narrow track such as "honors," "college-bound," "regular," or "remedial." Most labs were developed for heterogeneous groups that include some strong students, but you should be able to find plenty in here to use in just about any classroom.

This is not a textbook, and I do not claim that it offers a comprehensive treatment of secondary-school geometry. However, it addresses many essential ideas and can be a substantial part of math classes at many levels:

- *Middle school teachers* will find many labs that help prepare students for high school geometry by getting them to think visually and to become familiar with fundamental concepts, figures, and vocabulary. It is unfortunate that much of the curriculum currently available for middle school geometry consists of rote activities built around a few results when there is a wealth of possibilities.
- *Teachers of geometry* courses, whether traditional, inductive, or technology-based, will find many labs that approach key topics in their curriculum from a different point of view. In some cases, you will find, as I have, that the lessons in this book can replace the corresponding ones in your textbook. Other labs can be used to *preview* or *review* material that you teach in more traditional ways.
- *Teachers of integrated mathematics* courses at any secondary-school level will have no trouble finding suitable labs in this book, either to support the geometry component of their courses, or to enrich it.
- Finally, *trigonometry teachers* will find interesting material in Section 11. The approach I pursue there should make trigonometry material accessible to younger students.

Here is a rough estimate of grade levels appropriate for sections of this book:

Sections 1–8: grades 7–11

Sections 9–11: grades 9 and up

In some cases, I included lessons whose purpose is to introduce an idea that is a prerequisite to a subsequent lab. While these can usually be skipped in a class that is also using a geometry textbook, they are included for the use of middle school teachers or for teachers of integrated mathematics courses whose textbooks sometimes do not include those concepts. Such labs are designated as "getting ready" activities in the Teacher Notes.

Geometry and Proof

Formal proof has a central role in high school mathematics. Traditionally, proof has been introduced in the geometry course, but, unfortunately, this has not worked as well as many of us would like. In many traditional courses, the first proofs are of self-evident results like "the angle bisector divides the angle into equal angles," which is a sure way to baffle beginners. In other courses, proof is completely avoided.

A successful introduction to proof has to be rooted in rich mathematical content and in the use of discourse and reasoning. This book provides an enormous supply of rich mathematical content and opportunities for discourse and reasoning. While it does not take the next step of helping students learn to write formal proofs, it does set the stage for such learning. At the Urban School, we introduce formal proofs in the second semester of the course, building on the foundation laid by many of the labs in this book.

Still, important as it is, I don't see the introduction of proof as the only reason for teaching geometry:

- There is plenty of geometry content that is of great importance to further work in mathematics. I am thinking of topics such as measurement, distance and the Pythagorean Theorem, and similarity and scaling, all covered in the last four sections of this book.
- Geometry is surely an area in which the aesthetic appeal of mathematics is most clearly evident. I certainly hope that you will find time for the sections on symmetry and tiling, which will appeal to the artist in each of your students.
- Finally, geometric puzzles have a fascination for many. I have mined recreational mathematics for material that is both entertaining and educational, and I have inserted puzzles and other visual challenges (many of them original) throughout this book.

Using Manipulatives

Almost all the activities in this book are based on the use of manipulatives. This is not because I believe that there is some sort of magic that guarantees learning as soon as students are manipulating something. I have been involved with manipulatives in math education for over 25 years and have no such illusions. Still, I do believe that manipulatives can help in the following ways:

- Manipulatives can motivate students, particularly if they are used in a thought-provoking, puzzle-like way. In this book, I have tried to choose activities that are intrinsically interesting and that appeal to a range of students.
- Manipulatives can be the spark for significant discussions, both at the small-group level and with the whole class. In addition to the exercises on the worksheets, most labs also include Discussion questions that can lead to stronger understanding for all.

To get the greatest mathematical payoff from the use of the manipulatives, it is essential that you make explicit connections between what is learned in this context and the broader goals you have in your course. I have included suggestions for this in the Teacher Notes that accompany each lab.

Using the Discussion Questions

The Discussion questions tend to be more difficult, more general, or deeper in some ways than the rest of the lab. Generally, they are intended to follow the lab and can be used to spark whole-class or small-group conversation. However, be flexible and alert. Often, the Discussion questions can be used smack in the middle of the lab, when students are ready to think and talk about them. In many cases, the Discussion questions make excellent prompts for in-class or homework writing assignments. Written responses can serve as the basis for a richer discussion or as a valid assessment tool.

During most labs you should help students mostly with hints and nudges and encourage them to seek help from each other. However, it is likely that you will need to play a more directive and assertive role in the discussions. Do not be afraid to explain difficult ideas to students—once they know what the questions are. One way to make a lecture ineffective is to try to answer questions students don't have. The labs and Discussion questions that follow will generally raise the appropriate questions and prepare students to hear your explanations.

Using the Teacher Notes

The Teacher Notes are designed to help you prepare for the labs. They start with a list of prerequisite concepts, if there are any. This list can be crucial because you may be choosing individual lessons here and there rather than using the labs sequentially from beginning to end. The prerequisites are either general geometric ideas that are probably already a part of your course, or they are a specific lab or labs from an earlier part of this book.

The Teacher Notes also provide help with timing issues. In the never-ending conflict between "covering" and "discovering," you need to keep a balance and avoid going too far in either direction. Since, for most of us, the former has a tendency to dominate our thinking, I would encourage you to not rush through the labs. Most labs should fit in one class period, but for those that don't, it is better to allow two periods or use longer periods if you have those somewhere in your schedule, even if the consequence is that you have to limit yourself to doing fewer labs. I have tried to indicate the labs that are likely to take a long time, so you can plan accordingly. Alas, it is often the more valuable labs that take the most time.

One more thought about time: Be aware that getting to closure on all the Discussion questions could lengthen some labs more than you can afford in a single period, so you will have to choose which of those questions you want to address in a thorough manner. On rare occasions, there are questions in the body of the labs that are potentially very difficult or time-consuming. I tried to flag those in the Notes and Answers.

Tools and Concepts Overview

This table shows which sections use which tools, to teach which concepts. Use it in deciding what to teach, what manipulatives you will need, and so on.

"Shape" includes the definition, recognition, and understanding of various geometric shapes, plus concepts such as convexity and congruence. "Measuring" includes perimeter, area, surface area, and volume, as well as distance and square root.

	Pattern Blocks	Template	CircleTrig Geoboard	Tangrams	Cubes	Mirrors	11 × 11 Geoboard
Angles	1, 3, 5, 7	1, 3, 5, 6, 7	1, 3, 5				11
Shape	3	6, 7	3	2	4		6
Symmetry	2, 5	2, 5, 6	5		4	5	
Dimension		6			4, 10		
Tiling	7	7			7		
Measuring	9	9	9	10	8		8, 9
Similarity			11	10	10		10, 11
Trigonometry		11	11				

Some tools are always expected to be available. These include a calculator, a compass, a ruler, a protractor, and assorted types of papers: unlined paper, graph paper, and grid papers and record sheets you can duplicate from the back of this book.

If you have access to computers, I strongly recommend that you make use of them in teaching geometry. Programming languages with turtle graphics, such as Logo and Boxer, and interactive geometric construction programs, such as The Geometer's Sketchpad® and Cabri II™, offer many opportunities for great labs. This does not conflict with the "lower" technology of manipulatives: Each type of lab has its uses and its place in a full geometry program.

Specific Notes on Some of the Manipulatives

I recommend you buy the complete kit designed for this book. Two items in the kit—the CircleTrig Geoboard and the drawing template—are unique. If you already have a collection of commonly used manipulatives, you may get by with what you have, perhaps supplemented by a separate purchase of the unique items. Study this list to see what you'll need:

Pattern Blocks: Pattern blocks are very popular with students and pretty much indispensable to key labs in this book. They were invented in the 1960s by the Elementary Science Studies and can be found in most elementary schools. My students sometimes affectionately call them "kindergarten blocks," and no, they do not find them offensive.

Template: The *Geometry Labs* drawing template will allow your students to conveniently draw pattern blocks, regular polygons, and a decent selection of triangles and quadrilaterals. In addition, the fact that it incorporates a straightedge, an inch ruler, a centimeter ruler, a protractor, and a circle shape makes it a useful tool to have at all times in math class. If you already have other templates that fulfill some or all of those functions, you may be able to use those instead. The first time you use the template, ask students to trace all the figures on a piece of unlined paper to serve as a Template Reference Sheet. Every time you use a new figure in a lab, have students label the figure on the sheet.

CircleTrig Geoboard: The CircleTrig Geoboard I designed for *Geometry Labs* has pegs every 15 degrees around the circle and every 2 cm around the perimeter of the circumscribing square. The engraved degree marks on the circle and millimeter marks on two sides of the square make it the most versatile circle geoboard around, extending its use to trigonometry topics. The 11×11 square geoboard on the other side of the board makes this the best general-purpose geoboard you'll find. You can use a classic circle geoboard, which is much smaller but adequate for Sections 1–10. For Section 11, you're better off with the CircleTrig board, but you can survive by duplicating page 245.

Tangrams: Tangrams are a classic geometric puzzle, with a long history in and out of the classroom. They are used only in two sections, Sections 2 and 10.

Cubes: Cubes are useful as multi-purpose tools in a math classroom. If you have cubes other than the ones supplied in the *Geometry Labs* kit, you probably can use them for most labs that require cubes. Interlocking cubes are definitely preferable to non-interlocking cubes, and cubes that attach on all six faces are preferable to those that don't.

Mirrors: Mirrors are used only in Section 5. If you have other mirrors or some equivalent, those will work just as well.

11 × 11 Geoboard: Crucial in the latter half of the book, this geoboard makes an important connection with algebra and the Cartesian plane. The 5 × 5 geoboard is not really adequate for the purposes of these labs, but dot paper is a passable substitute.

Related Publications

More labs along the lines of the ones in this book can be found in the *Pentominoes* and *Super Tangrams*™ books, which I created for Creative Publications. They are suitable for both the middle school and high school level.

1 Angles

An essential foundation of secondary school geometry is the concept of angle. Many students do not have enough experience with angles when, at the beginning of a geometry class, we tell them how to use a protractor. Often, we are surprised at how difficult this idea is for many students. I have banged my head against that wall, repeating over and over to some students how a protractor is used, and eventually made some headway. But that was a frustrating approach. What I was missing was that students have trouble because it is difficult to measure something when you are not sure what it is. I developed the first lessons in this section in order to build the intuitive foundation necessary for much of the work in a traditional geometry course. Of course, understanding angles is also necessary in reform-minded courses (such as the one I teach at the Urban School of San Francisco) and in later work in trigonometry.

Many geometry textbook authors are concerned about distinguishing angles from their measures, and use a different notation for each. This distinction is rather subtle and, frankly, I do not burden my students with it. There is not much to say about angles unless we are talking about their measures. If you are concerned with inconsistencies between sheets copied from this book and the textbook you use, tell your students that notation and language vary from book to book. It is unlikely to upset them, as long as you make clear what notation you want them to use in tests and other assessments.

I do not believe in the widespread but absurd restriction that angle measurements must be less than 180°. We are talking to some students who can turn 360° on their skateboards, and to some who will soon be studying trigonometric functions. All of them live in a world where polygons can be nonconvex (consider the Star Trek® insignia), and therefore include angles greater than 180°. Angles whose measures are between 180° and 360° are called *reflex* angles. You may add this word to your students' vocabulary, along with the better-known *acute, right, obtuse,* and *straight,* even if the word is not in the textbook you use. One of the theorems we commonly teach in geometry (and in this book) is about the sum of the angles in a polygon. The theorem still holds in the case of nonconvex polygons. Why limit this result with the artificial constraint that angles cannot measure more than 180°?

One particular aspect of the study of angles is the study of central and inscribed angles in a circle. This is an interesting topic, but difficult, and is usually relegated to the very end of textbooks. Consequently, some of us never get to it, and the designers of standardized tests use this fact to separate students in honors classes from the rest of the population. The activities in the second part of this section are intended to give students in all sorts of geometry classes a jump-start on the ideas involved. You may teach these early in the year and review them later when you reach that point in your textbook; alternatively, you may save them for later and teach them right before they arise in the textbook. However, if there is a risk that you may not get to that point, you may consider moving that whole topic to an earlier time in your course plan.

I ended this section with a related topic: "soccer angles." These lessons are introduced with a motivating real-world problem and lead to a tough optimization challenge: Where is the best place on the field to shoot at the goal from?

See page 167 for teacher notes to this section.

LAB 1.1
Angles Around a Point

Name(s) ______________________

Equipment: Pattern blocks

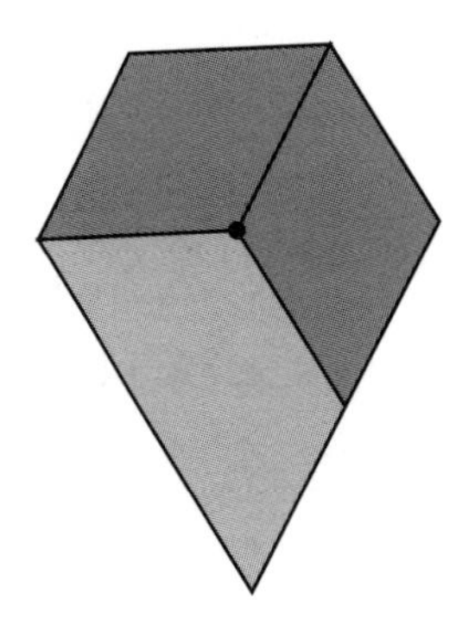

Place pattern blocks around a point so that a vertex (corner) of each block touches the point and no space is left between the blocks. The angles around the point should add up to exactly 360°.

For example, with two colors and three blocks you can make the figure at right.

Use the chart below to keep track of your findings.

- Every time you find a new combination, circle the appropriate number on the list below.
- Cross out any number you know is impossible.
- If you find a possible number that is not on the list, add it.

Since the two-colors, three-blocks solution is shown above, it is circled for you.

Colors:	**How many blocks you used:**									
all blue	3	4	5	6						
all green	3	4	5	6						
all orange	3	4	5	6						
all red	3	4	5	6						
all tan	3	4	5	6						
all yellow	3	4	5	6						
two colors	③	4	5	6	7	8	9	10	11	12
three colors	3	4	5	6	7	8	9	10	11	12
four colors	3	4	5	6	7	8	9	10	11	12
five colors	3	4	5	6	7	8	9	10	11	12
six colors	3	4	5	6	7	8	9	10	11	12

How many solutions are there altogether? __________

Discussion

A. Which blocks offer only a unique solution? Why?

B. Why are the tan block solutions only multiples of 4?

C. Explain why the blue and red blocks are interchangeable for the purposes of this activity.

D. Describe any systematic ways you came up with to fill in the bottom half of the chart.

E. How do you know that you have found every possible solution?

F. Which two- and three-color puzzles are impossible, and why?

G. Which four-color puzzles are impossible, and why?

H. Why is the five-color, eight-block puzzle impossible?

I. Which six-color puzzles are impossible, and why?

LAB 1.2

Angle Measurement

Name(s) ____________________

Equipment: Pattern blocks, template

1. What are the measures of the angles that share a vertex at the center of
   - **a.** A Chrysler symbol? ___________
   - **b.** A Mercedes symbol? ___________
   - **c.** A peace sign? ___________
   - **d.** A clock, between consecutive hours? ___________
   - **e.** A cross? ___________

2. Find the measures of all the angles for each of the pattern blocks shown below. Write the angle measures in these shapes.

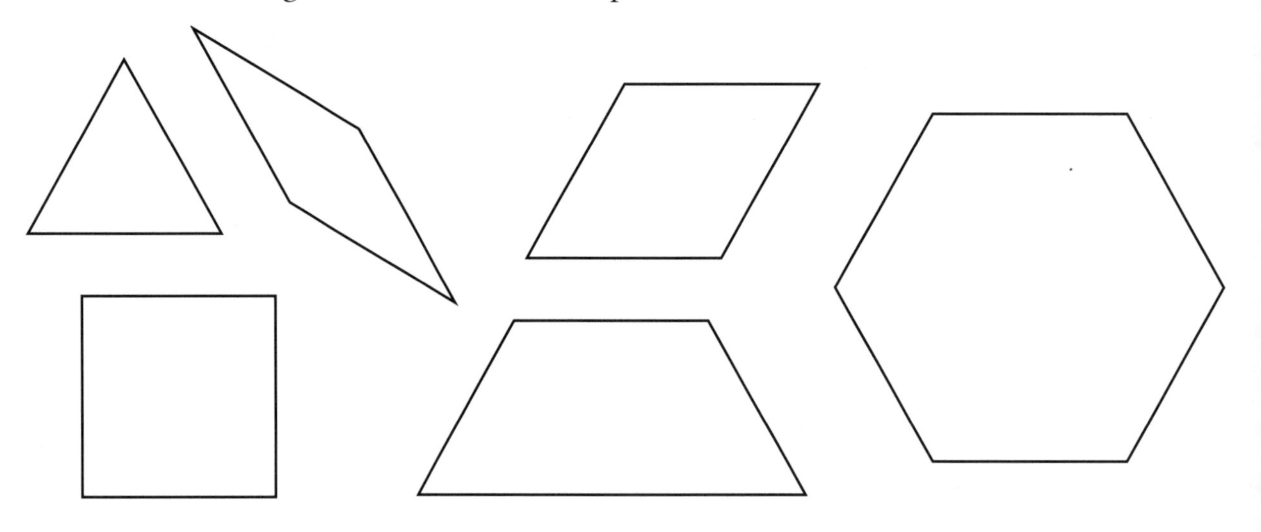

3. One way to measure angles is to place smaller angles inside larger ones. For example, six copies of the small angle on the tan pattern block fit inside the figure below. This figure, called a **protractor**, can be used to measure all pattern block angles.
   - **a.** Mark the rest of the lines in the figure with numbers.
   - **b.** Use it to check the measurements of the pattern block angles.
   - **c.** Using the tan pattern block, add the 15° lines between the 30° lines shown on the protractor.

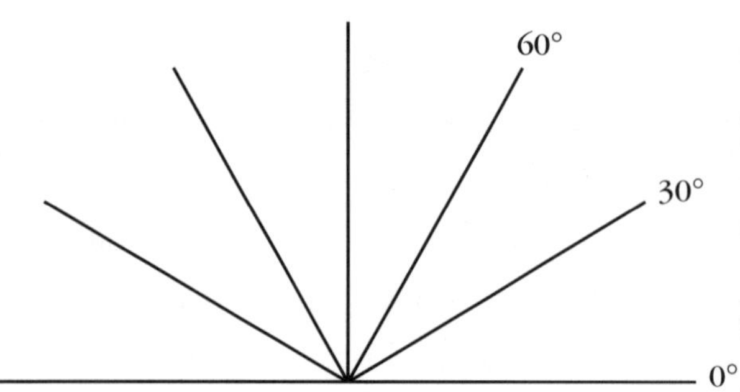

4. Use the protractor on your template to measure the angles in the triangles below. For each one, add up the angles. Write the angle sum inside each triangle.

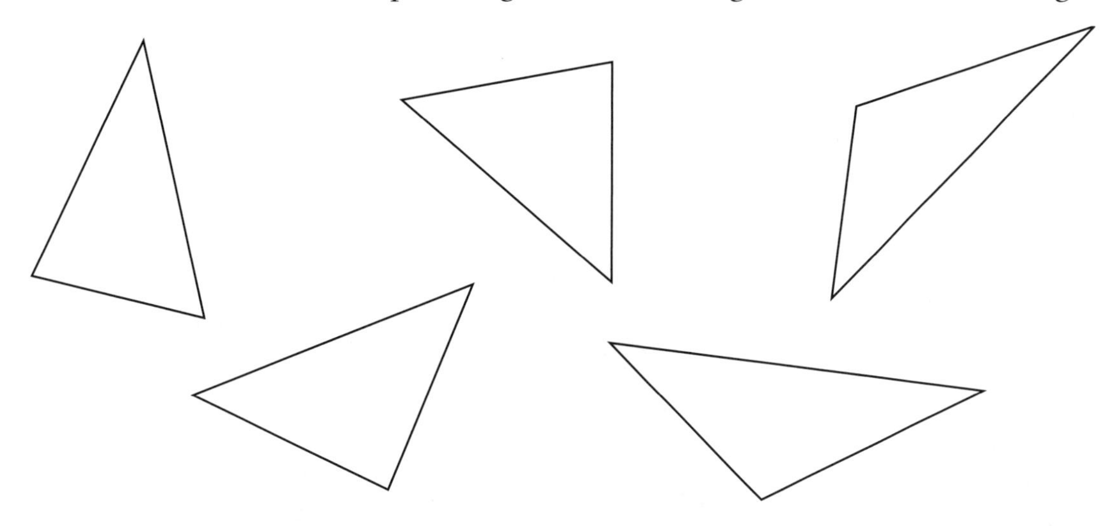

5. Use the protractor on your template to draw the following angles in the space below.

a. 20°

b. 50°

c. 100°

Discussion

Which of the questions in Problem 1 did you find most difficult to answer? What made the others easier?

LAB 1.3
Clock Angles

Name(s) ______________________

What angles do the hour and minute hands of a clock make with each other at different times? Write an illustrated report.

- Start by figuring it out on the hour (for example, at 5:00).
- Then see if you can answer the question for the half-hour (for example, at 5:30).
- Continue exploring increasingly difficult cases.

Remember that the hour hand moves, so, for example, the angle at 3:30 is *not* 90° but a little less, since the hour hand has moved halfway toward the 4.

Use the clocks below for practice. Cut and paste clocks into your report.

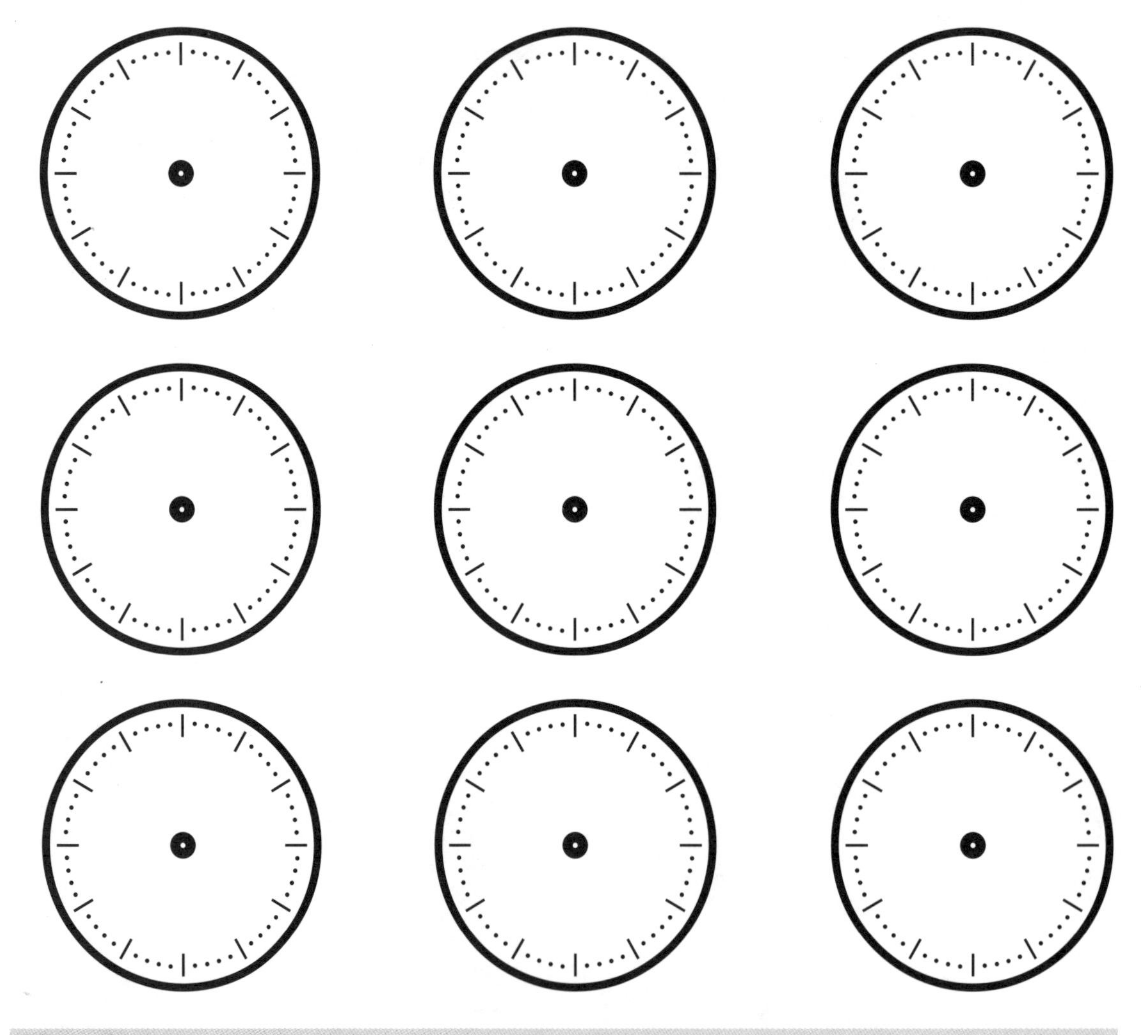

Discussion

A. How many degrees does the hour hand travel in an hour? In a minute?

B. How many degrees does the minute hand travel in an hour? In a minute?

AB 1.4

Name(s) ______________________

Angles of Pattern Block Polygons

■ **Equipment:** Pattern blocks

1. Find the *sum* of the angles for each pattern block, and record them below.
   a. Green __________
   b. Orange __________
   c. Blue __________
   d. Tan __________
   e. Red __________
   f. Yellow __________
2. Which blocks have the same sum?
3. Use pattern blocks to make a polygon such that the sum of its angles is the same as the sum for the hexagon. Sketch your solution in the space at right. How many sides does it have?
4. Use pattern blocks to make a polygon such that the sum of its angles is less than the sum for the hexagon but more than the sum for the square. Sketch your solution in the space at right. How many sides does it have?
5. Use pattern blocks to make a polygon such that the sum of its angles is greater than the sum for the hexagon. Sketch your solution in the space at right. How many sides does it have?
6. For each number of sides from 3 to 12, make a pattern block polygon and find the sum of its angles. Sketch your solutions on a separate sheet of paper and fill out the table below.

Sides	Sum of the angles
3	
4	
5	
6	
7	

Sides	Sum of the angles
8	
9	
10	
11	
12	

7. If a pattern block polygon had 20 sides, what would be the sum of its angles? Explain.

8. If the sum of the angles of a pattern block polygon were 4140°, how many sides would it have? Explain.

9. If the sum of the angles of a pattern block polygon were 450°, how many sides would it have? Explain.

10. What is the relationship between the number of sides of a pattern block polygon and the sum of its angles? Write a sentence or two to describe the relationship, or give a formula.

LAB 1.5
Angles in a Triangle

Name(s) ______________________

■ **Equipment:** Template

Types of triangles

Obtuse: contains one obtuse angle
Right: contains one right angle
Acute: all angles are acute

1. Could you have a triangle with two right angles? With two obtuse angles? Explain.

2. For each type of triangle listed below, give two possible sets of three angles. (In some cases, there is only one possibility.)

   a. Equilateral: ______________, ______________

   b. Acute isosceles: ______________, ______________

   c. Right isosceles: ______________, ______________

   d. Obtuse isosceles: ______________, ______________

   e. Acute scalene: ______________, ______________

   f. Right scalene: ______________, ______________

   g. Obtuse scalene: ______________, ______________

3. If you cut an equilateral triangle exactly in half, into two triangles, what are the angles of the "half-equilateral" triangles? ______________

4. Which triangle could be called "half-square"? ______________

5. Explain why the following triangles are impossible.

   a. Right equilateral

   b. Obtuse equilateral

 Name(s) ____________________

Angles in a Triangle (continued)

6. Among the triangles listed in Problem 2, which have a pair of angles that add up to 90°?

7. Make up two more examples of triangles in which two of the angles add up to 90°. For each example, give the measures of all three angles.

______________, ______________

8. Complete the sentence:
"In a right triangle, the two acute angles . . . "

9. Trace all the triangles on the template in the space below and label them by type: equilateral (EQ), acute isosceles (AI), right isosceles (RI), obtuse isosceles (OI), acute scalene (AS), right scalene (RS), half-equilateral (HE), obtuse scalene (OS).

LAB 1.6

Name(s) ____________________

The Exterior Angle Theorem

In this figure, $\angle A = 60°$ and $\angle B = 40°$. One side of $\angle C$ has been extended, creating an *exterior angle:*

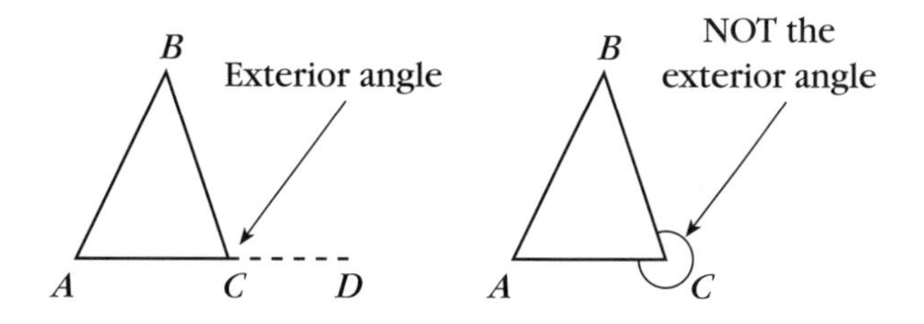

1. Without measuring, since this figure is not to scale, find the interior angle at C ($\angle ACB$) and the exterior angle at C ($\angle BCD$).

   Interior angle at C __________

   Exterior angle at C __________

2. Make up two other examples of triangles where two of the angles add up to 100°. Write all three angles for each triangle in the spaces below. Sketch the triangles in the space at right.

   ______________, ______________

3. For each of your triangles, find the measures of all three exterior angles.

   Triangle one: ______________

   Triangle two: ______________

4. In the space at right, sketch an example of a triangle ABC where the exterior angle at B is 65°.

   a. What is the measure of the interior angle at B?

   ____

   b. What is the sum of the other two interior angles ($\angle A$ and $\angle C$)?

   ____

5. Repeat Problem 4 with another example of a triangle ABC with a 65° exterior angle at B.

   a. What is the interior angle at B?

   ____

   b. What is the sum of the other two interior angles ($\angle A$ and $\angle C$)?

   ____

Name(s) ____________________

6. I am thinking of a triangle *ABC*. Exterior angle *A* is 123°. Interior angle *B* is given below. What is the *sum* of interior angles *B* and *C*? (Use a sketch to help you work this out.)

a. If $\angle B = 10°$ __________

b. If $\angle B = 20°$ __________

c. Does it matter what $\angle B$ is? Explain.

7. I am thinking of a triangle *ABC*. Exterior angle *A* is $x°$. What is the sum of interior angles *B* and *C*? Explain how you get your answer.

8. Complete this statement and explain why you think it is correct:

The Exterior Angle Theorem: An exterior angle of a triangle is always equal to . . .

9. What is the sum of the two acute angles in a right triangle? Is this consistent with the exterior angle theorem? Explain. (**Hint:** What is the exterior angle at the right angle?)

10. What is the sum of all three exterior angles of a triangle? Find out in several examples, such as the ones in Problems 1, 2, 4, and 5. Explain why the answer is always the same.

Discussion

A. How many exterior angles does a triangle have?

B. What is the sum of an interior angle and the corresponding exterior angle?

C. Draw a triangle, and extend all the sides in both directions. Mark the interior angles with an *i,* the exterior angles with an *e,* and the angles that are vertical to the interior angles with a *v*.

D. In Problems 4–6, are the triangles discussed acute, right, or obtuse? How do you know?

E. If a triangle ABC has an exterior angle A of 50° and an interior angle B of $x°$, what is the interior angle C in terms of x? What is the sum of angles B and C in terms of x?

LAB 1.7

Name(s) ______________________

Angles and Triangles in a Circle

■ **Equipment:** Circle geoboard, Circle Geoboard Paper

Types of triangles

Equilateral (EQ)
Right isosceles (RI)
Acute scalene (AS)
Half-equilateral (HE)
Acute isosceles (AI)
Obtuse isosceles (OI)
Right scalene (RS)
Obtuse scalene (OS)

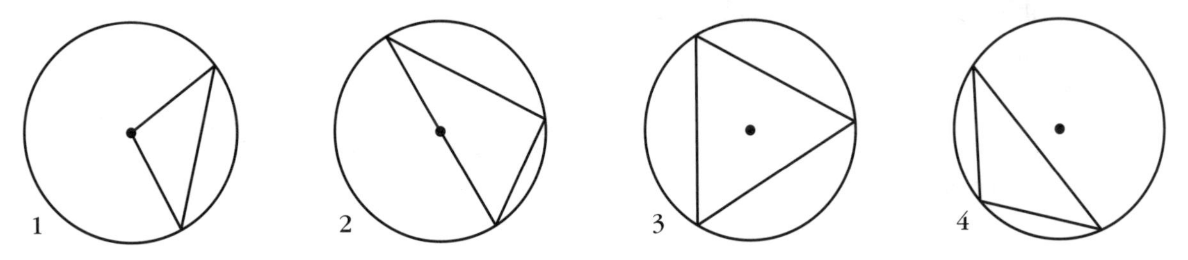

1. Make triangles on the circle geoboard, *with one vertex at the center* and the other two on the circle. (See circle 1 above.)
   a. Make one of each of the eight types of triangles listed above, if possible. You do not have to do them in order!
   b. Sketch one of each of the types of triangles it was possible to make on circle geoboard paper. Identify which type of triangle it is, and label all three of its angles with their measures in degrees. (Do not use a protractor! Use geometry to figure out the angles.)
2. Thinking back:
   a. What is true of all the possible triangles in Problem 1?

   b. Summarize your strategy for finding the angles. Describe any shortcuts or formulas you used.

Definition: A triangle is *inscribed* in a circle if all three of its vertices are on the circle.

3. Repeat Problem 1 with inscribed triangles such that the circle's center is *on a side* of the triangle. (See circle 2.) **Hint:** Drawing an additional radius should help you find the measures of the angles.

4. Thinking back:

a. Summarize your strategy for finding the angles in Problem 3. Describe any shortcuts or formulas.

b. Write the exterior angle theorem. How can it be used to help find the angles?

5. What is true of all the possible triangles in Problem 3? Prove your answer.

6. Repeat Problem 1 with inscribed triangles such that the circle's center is *inside* the triangle. (See circle 3.)

7. Repeat Problem 1 with inscribed triangles such that the circle's center is *outside* the triangle. (See circle 4.)

Discussion

A. Which triangles are impossible in Problem 1? Why?

B. In Problem 1, what is the smallest possible angle at the vertex that is at the circle's center? Explain.

C. Explain why the angle you found in Question B is the key to Problem 1.

D. How do you use the isosceles triangle theorem to find the other two angles of the triangles in Problem 1?

E. What is true of all the triangles in Problem 6? All the triangles in Problem 7? Explain.

LAB 1.8

Name(s) ______________________

The Intercepted Arc

Definitions:

A *central angle* is one with its vertex at the center of the circle.
An *inscribed angle* is one with its vertex on the circle.
The arc *intercepted* by an angle is the part of the circle that is inside the angle.
The *measure of an arc* in degrees is the measure of the corresponding central angle.

1. Use the definitions above to fill in the blanks: In the figure below left, $\angle APB$ is ____________, $\angle AOB$ is ____________, and $\widehat{AB}$ is ______________________________.

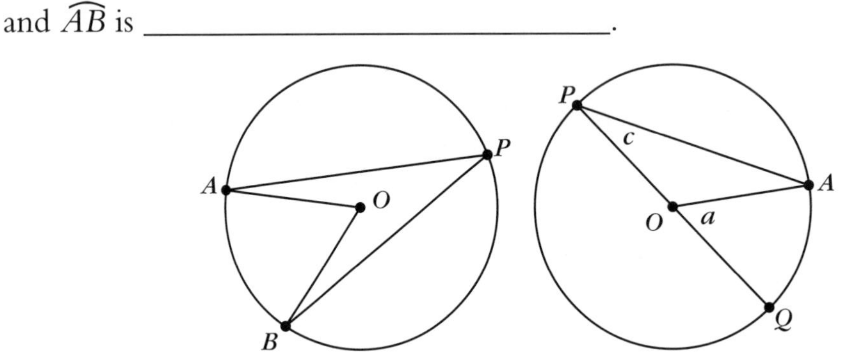

2. In the figure above right, which arc is intercepted by $\angle AOQ$? __________ Which arc is intercepted by $\angle APQ$? __________

3. If $\angle a = 50°$, what is $\angle c$? Explain.

4. In general, what is the relationship between $\angle a$ and $\angle c$? Explain.

5. In the figure at the right, if $\angle AOB = 140°$ and $\angle a = 50°$, what is $\angle b$? What is $\angle c$? What is $\angle d$? What is $\angle APB$?

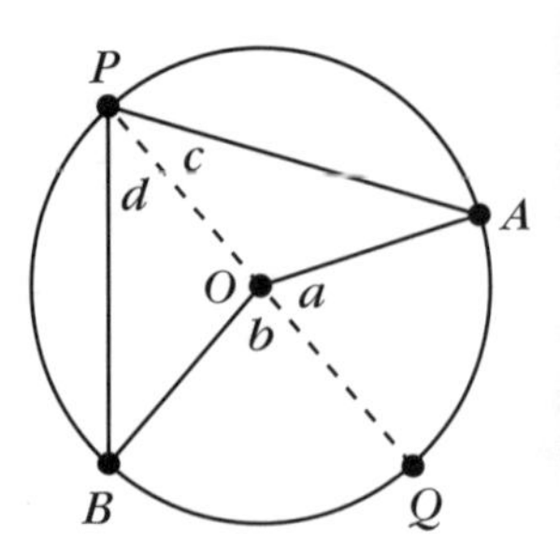

6. Repeat Problem 5 with $\angle AOB = 140°$ and three different values for $\angle a$. Arrange your results in a table.

Name(s) ____________________

7. What is the relationship between $\angle APB$ and $\angle AOB$? Explain.

8. Write a sentence about the relationship between an inscribed angle and the corresponding central angle. If you do this correctly, you have stated the inscribed angle theorem.

9. On a separate sheet, use algebra to prove the inscribed angle theorem for the case illustrated in Problem 5.

10. There is another case for the figure in Problem 5, where O is outside of $\angle APB$. On a separate sheet, draw a figure and write the proof for that case.

Discussion

A. Find the measure of angles a, b, c, and d in the figure at right. Explain.

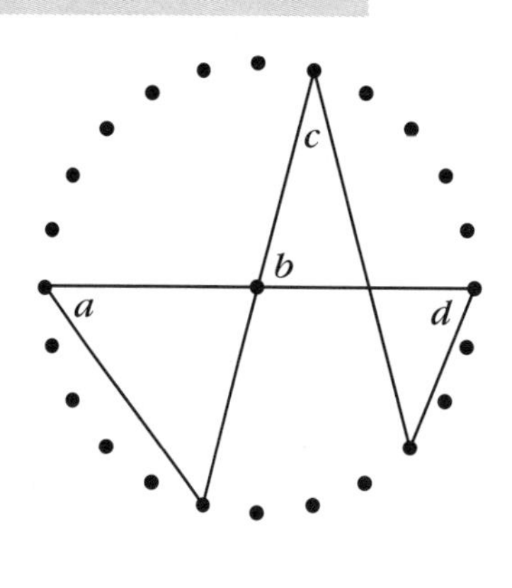

B. What are the interior angles in the triangles below?

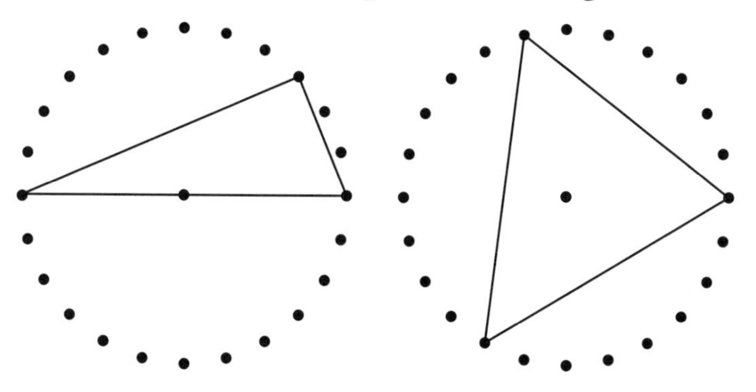

C. What are the measures of the central angle, the corresponding inscribed angle, and the intercepted arc in the figure at right?

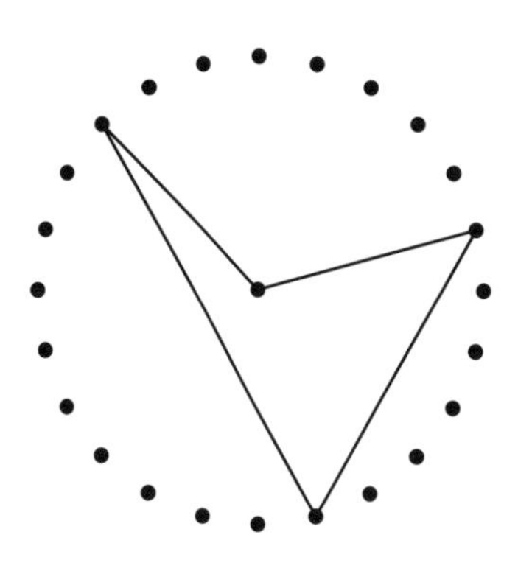

D. Inscribe a triangle in the circle geoboard so that its interior angles are 45°, 60°, and 75°.

LAB 1.9

Name(s) ____________________

Tangents and Inscribed Angles

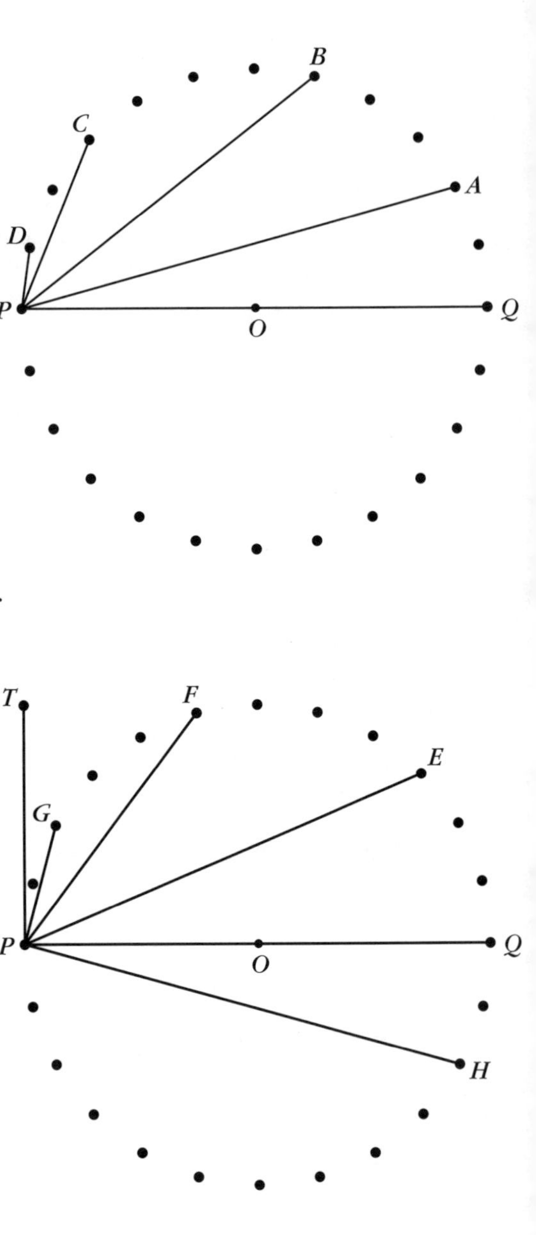

1. In the figure at right, find the intercepted arc and the measure for each of the following angles.

   a. $\angle QPA$ intercepts arc __________.

   $\angle QPA =$ __________

   b. $\angle QPB$ intercepts arc __________.

   $\angle QPB =$ __________

   c. $\angle QPC$ intercepts arc __________.

   $\angle QPC =$ __________

   d. $\angle QPD$ intercepts arc __________.

   $\angle QPD =$ __________

   e. Explain how you found the angle measures.

2. Segment PT is *tangent* to the circle (it touches it in exactly one point, P, which is called the *point of tangency*).

   a. What arc is intercepted by $\angle QPT$?

   ____

   b. What is the measure of $\angle QPT$?

   ____

3. **Important:** A segment that is tangent to a circle is ________________ to the radius at the point of tangency.

4. In the figure above, $\overline{PT}$ is tangent to the circle. Find the intercepted arc and the measure for each of the following angles. Explain on a separate sheet how you found the angle measures.

   a. $\angle TPE$ intercepts arc __________.

   $\angle TPE =$ __________

   b. $\angle TPF$ intercepts arc __________.

   $\angle TPF =$ __________

   c. $\angle TPG$ intercepts arc __________.

   $\angle TPG =$ __________

   d. $\angle TPH$ intercepts arc __________.

   $\angle TPH =$ __________

5. Use Problem 4 to check that the inscribed angle theorem still works if one side of the angle is tangent to the circle.

LAB 1.10

Name(s) ______________________

Soccer Angles

■ **Equipment:** Cardboard, unlined paper, Soccer Angles Worksheet, Soccer Circles Worksheet, scissors, straight pins

Soccer goals are 8 yards wide. Depending on where a player is standing, the angle she makes with the goalposts could be larger or smaller. We will call this angle the *shooting angle*. In this figure, the space between G and H is the goal, P is the player, and $\angle GPH$ is the shooting angle.

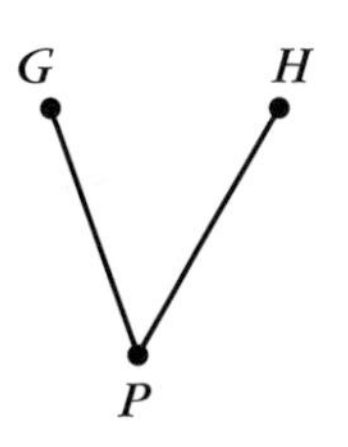

1. Place a piece of paper on a cardboard backing. Place two pins as goalposts two inches apart, centered, near the top of the page. Label the goalposts G and H.
2. Cut out these angles from the Soccer Angles Worksheet: 20°, 30°, 40°, 60°, 90°, 120°.
3. Using the paper angle and the pins, find the location of all points that are the vertices of a shooting angle equal to 40°.
4. Repeat Problem 3 with the other five angles cut from the Soccer Angles Worksheet, indicating clearly which points correspond to each shooting angle.

If you did this correctly, you should have found that the locations form arcs of circles that pass through G and H. You may remove the pins and ask your teacher for the Soccer Circles Worksheet, where the circles are drawn very accurately for you.

5. Label each arc with the measure of the corresponding shooting angle.
6. Which of the arcs is a half-circle? __________
7. The center of each of the six arcs is marked. Label each center so you will know which shooting angle it belongs to. (Use the notation C_{20}, C_{40}, and so on.)
8. Imagine a player is standing at C_{40}, the center of the 40° circle. What is the shooting angle there? ________________
9. Find the shooting angle for a person standing *at the center* of each of the remaining five circles. What is the pattern?

10. Find the center of the 45° circle without first finding points on the circle.

Discussion

A. The vertices of the 40° shooting angles all lie on a circle. They are the vertices of inscribed angles. Where is the arc that is intercepted by those angles?

B. How is the inscribed angle theorem helpful in understanding what happens in this activity?

For Questions C–E, refer to the Soccer Discussion Sheet.

C. Imagine a player is running in a direction parallel to the goal, for example, on line L_0. Where would he get the best (greatest) shooting angle? (Assume the player is practicing, and there are no other players on the field.)

D. Imagine a player is running in a direction perpendicular to the goal, on a line that intersects the goal, for example, L_2. Where would she get the best shooting angle?

E. Imagine you are running in a direction perpendicular to the goal, on a line that does *not* intersect the goal, such as L_3 or L_4. Where would you get the best shooting angle?

Soccer Angles (continued)

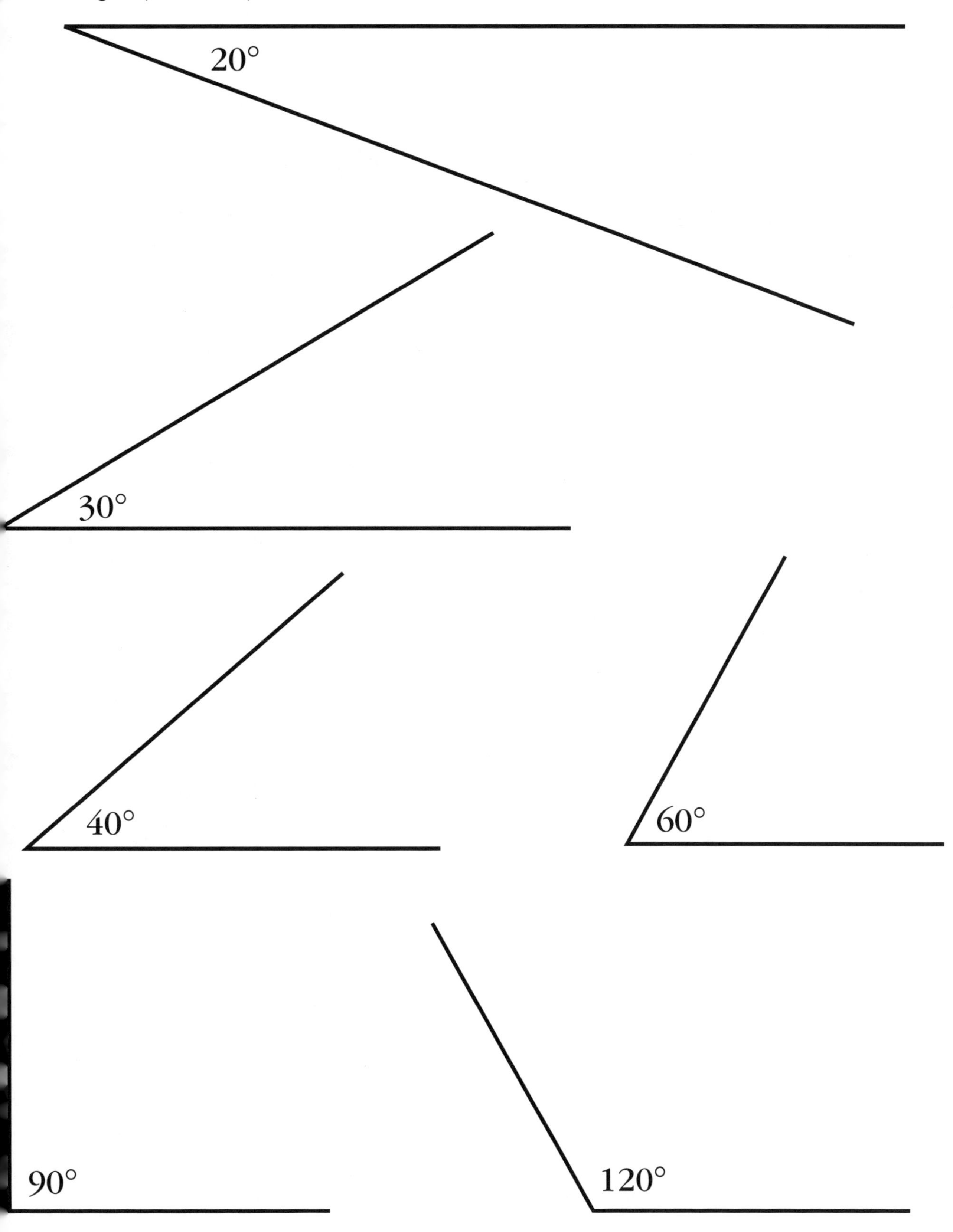

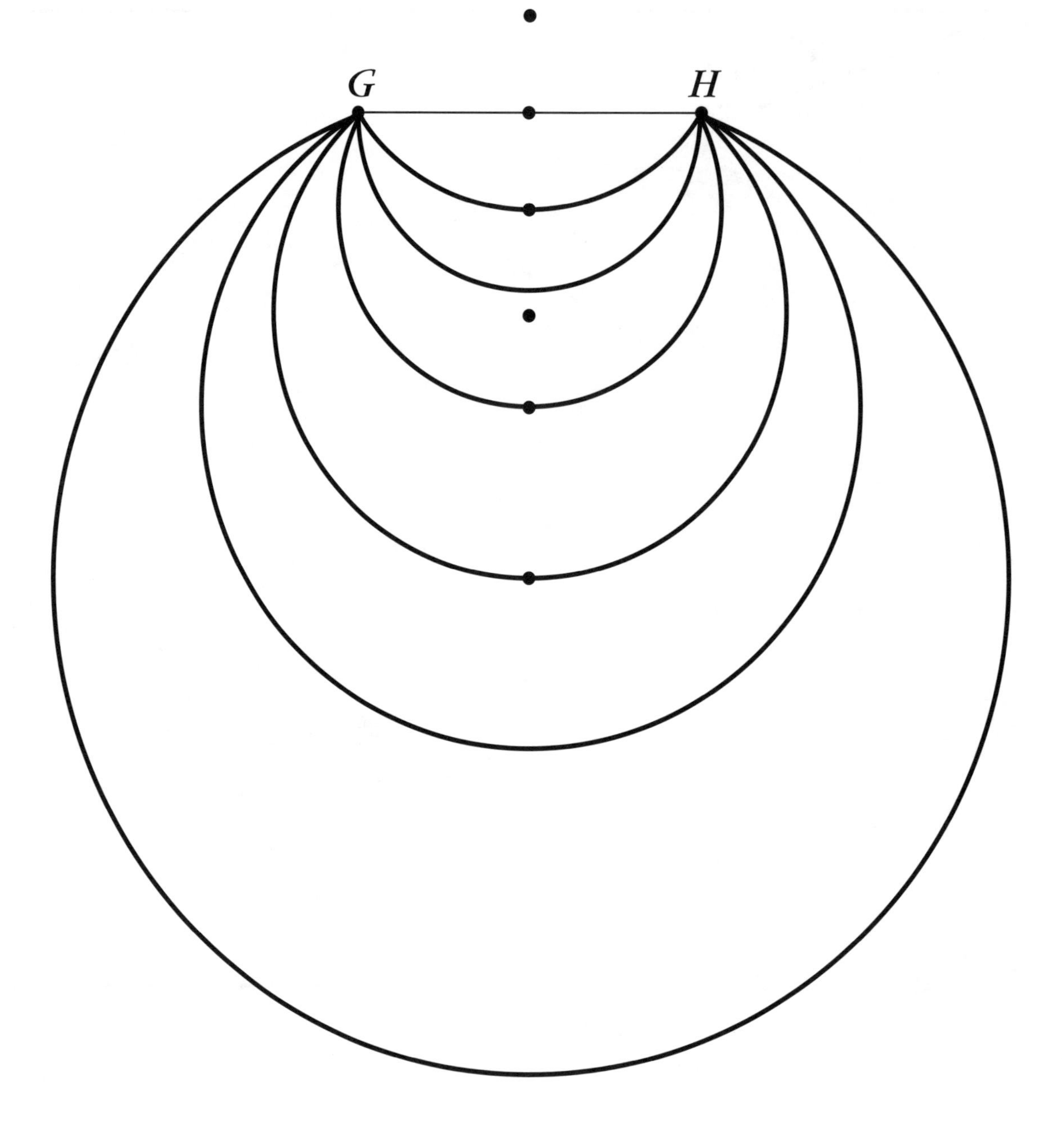
G
H

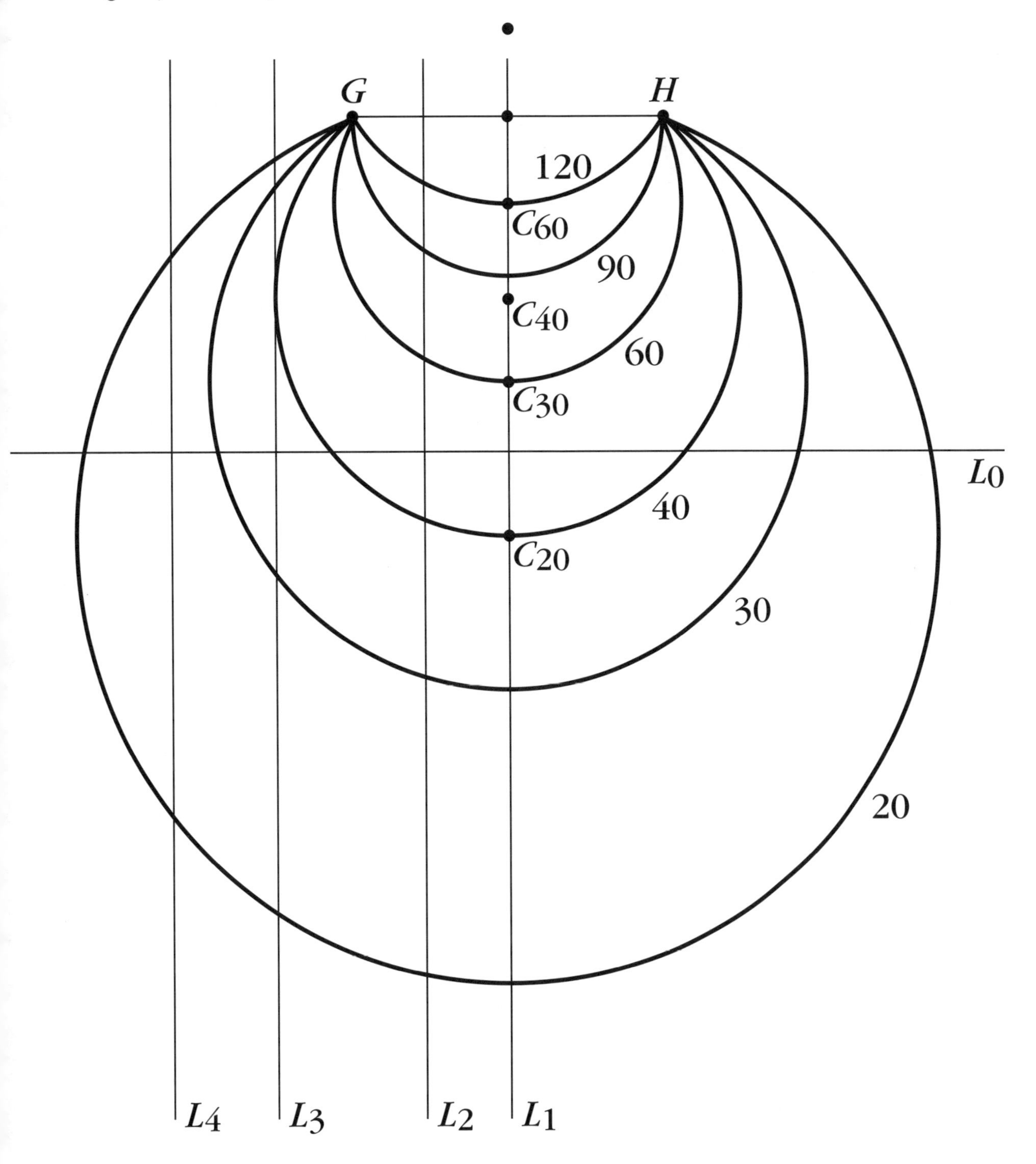
G
H
120
C_{60}
90
C_{40}
60
C_{30}
L_0
40
C_{20}
30
20
L_4
L_3
L_2
L_1

Soccer Goal Worksheet

G H

2 Tangrams

Tangram puzzles are quite accessible, and they help develop students' feel for the ubiquitous isosceles right triangle. This short section introduces some basic vocabulary and concepts about polygons in a context drawn from recreational mathematics. In addition, other concepts that will return in later sections make their first appearance here: square roots, similarity, symmetry, and convexity.

For more on tangrams, see:

Amusements in Mathematics, by Henry Dudeney (Dover Books)

Time Travel and Other Mathematical Bewilderments, by Martin Gardner (W. H. Freeman & Co.)

The latter book also features a bibliography about tangrams.

See page 173 for teacher notes to this section.

LAB 2.1
Meet the Tangrams

Name(s) ____________________

■ **Equipment:** Tangrams

1. How many tangrams are there per set? _________ Make sure you have a complete set.

2. The tangram figures below appeared in *Amusements in Mathematics,* a 1917 book by British puzzlesmith Henry Dudeney. Try making some of them. Record the ones you made.

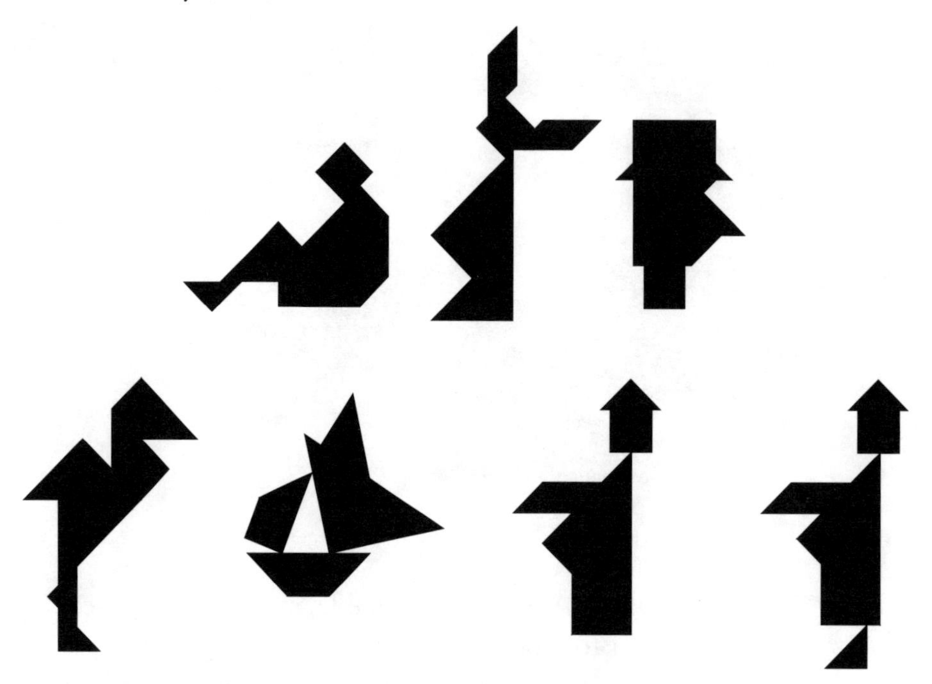

3. What interesting figures of your own can you make by combining all seven tangrams? Sketch them below and give them names.

4. What are the geometric names for each of the tangram shapes?

5. Trace each tangram shape. Flip it over. Does it still fit in the outline you traced?

6. What are the measures of the angles for each tangram shape? Find the angle measures without using a protractor, and write them down in the traced figures from Problem 5.

7. Write instructions you would give someone for making an accurate set of tangram pieces.

Discussion

A. How are the five tangram triangles related to one another?

B. How are the tangram triangles related to the other tangram shapes?

C. For each tangram, how much turning (how many degrees, or what fraction of a whole circle) do you have to do before it fits again in its traced outline?

D. Note that the last two figures shown on the previous page appear to be almost identical, except that one of them has a foot. Each of them was made with the whole set of seven tangrams. Where did the second man get his foot?

LAB 2.2

Name(s) ______________________

Tangram Measurements

- **Equipment:** Tangrams

Definition: The *hypotenuse* of a right triangle is the side opposite the right angle. The other two sides are the *legs*.

Fact: The hypotenuse of the small tangram triangle is exactly 2 inches in length.

1. Use that fact and logic to find as many of the other tangram side lengths as you can, as well as the area of each tangram piece. (Do not measure with a ruler!) If you get stuck, go on to the next piece, then come back to the ones you didn't get and try again. The rest of the activity will help you find measures you still can't get. Enter the length measurements along the sides of the reduced figures below, and write the areas inside the figures.

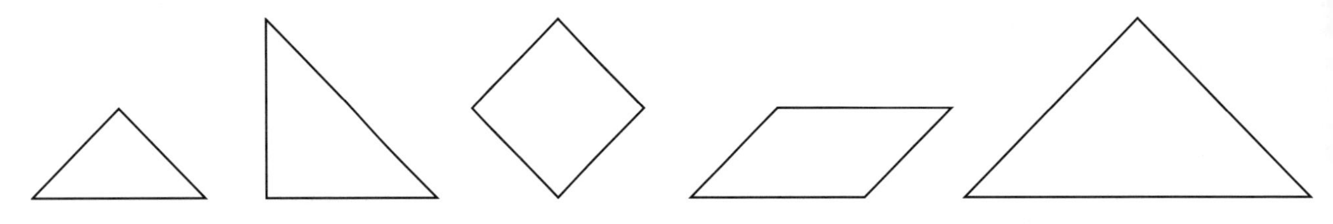

The rest of this lab will guide you through questions that should help you find any answers you did not find when working on Problem 1.

2. Cover the medium triangle with the two small triangles, then answer the following questions:

   a. How long is one leg of the medium tangram triangle? __________

   b. What is the area of the medium tangram triangle? __________

   c. What is the area of each small tangram triangle? __________

   d. What is the area of the tangram square? __________

Reminder: 49 is the square of 7, and 7 is the square root of 49.

3. Be careful when answering the following questions.

   a. If the side of a square is 9 units long, what is the square's area? __________

   b. If a square has an area of 9 square units, how long is one of its sides? __________

   c. If the side of a square is 5 units long, what is the square's area? __________

   d. If a square has an area of 5 square units, how long is one of its sides? __________

4. Generalize:
   a. If the side of a square is s units long, what is the square's area? __________
   b. If a square has an area of A square units, how long is one of its sides? __________
5. How long is the side of the tangram square? __________
6. Finish Problem 1.
7. Write an illustrated paragraph to explain how to obtain the measurements of all the tangram pieces from the hypotenuse of the small tangram triangle.

Discussion

A. In each of the tangram triangles, what is the ratio of the hypotenuse to the leg?

B. Each tangram triangle is a scaled version of each other tangram triangle. What is the *scaling factor* from the small to the medium? From the medium to the large? From the small to the large? (The scaling factor is the ratio of corresponding sides, *not* the ratio of areas.)

C. Make a square using the two large tangram triangles. What is its area? Express the length of its side in two different ways.

LAB 2.3
Tangram Polygons

Name(s) ____________________

- **Equipment:** Tangrams

- Make geometric figures, using any number of tangram pieces from one set.
- Keep track of your figures by checking boxes in the chart below.
- On the left side of the chart, add the names of other geometric figures you make, then check the box for the number of pieces you used in making them.

	How many pieces you used						
	1	2	3	4	5	6	7
Triangle							
Square							
Parallelogram							

Discussion

A. Which of the puzzles cannot be solved? For example, a one-piece, nonsquare rectangle is impossible. Are there others? Mark those puzzles with an X.

B. Explain why a six-piece square is impossible.

LAB 2.4
Symmetric Polygons

Name(s) ______________________

■ **Equipment:** Tangrams

This tangram polygon is not symmetric:

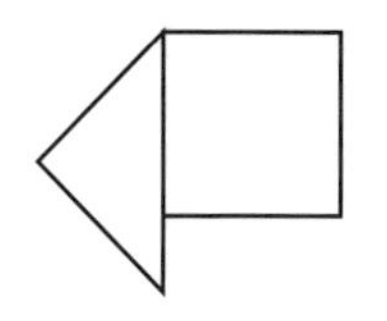

By adding a small tangram triangle, it can be made mirror symmetric:

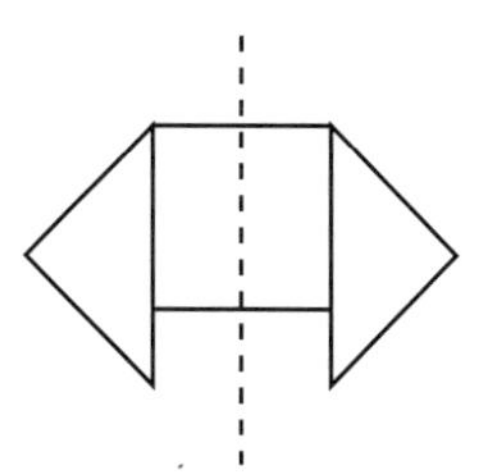

. . . or rotationally symmetric:

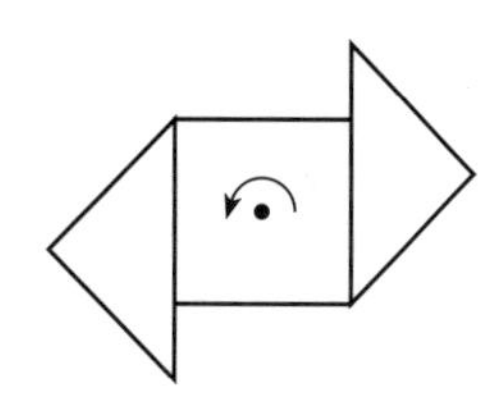

1. Add one or more tangram pieces to each tangram pair below to make mirror-symmetric polygons. Sketch your solutions on the figures.

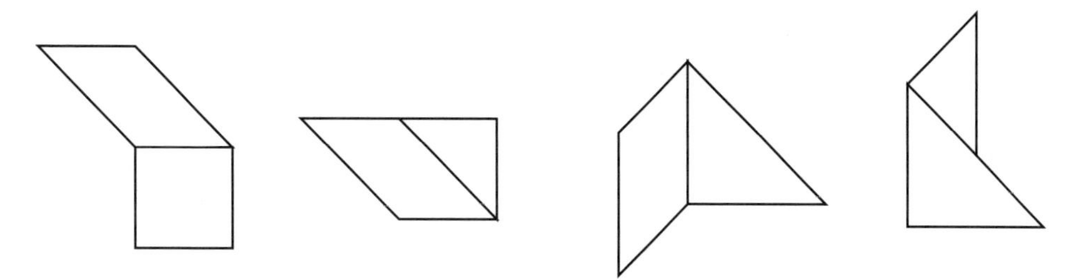

2. Repeat Problem 1, making rotationally symmetric polygons.

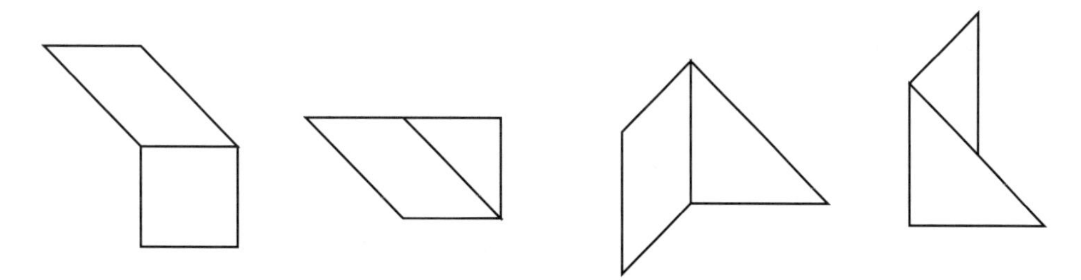

3. Create your own symmetric tangram polygons. Sketch them on the back of this sheet.

Discussion

A. In Problems 1 and 2, is it more elegant to create symmetric figures by adding as few additional pieces as possible, or as many additional pieces as possible?

B. Create your own puzzles in the style of Problems 1 and 2.

C. Is it possible for a tangram polygon to exhibit both mirror *and* rotational symmetry?

LAB 2.5
Convex Polygons

Name(s) ____________________

■ **Equipment:** Tangrams

The figures below are convex.

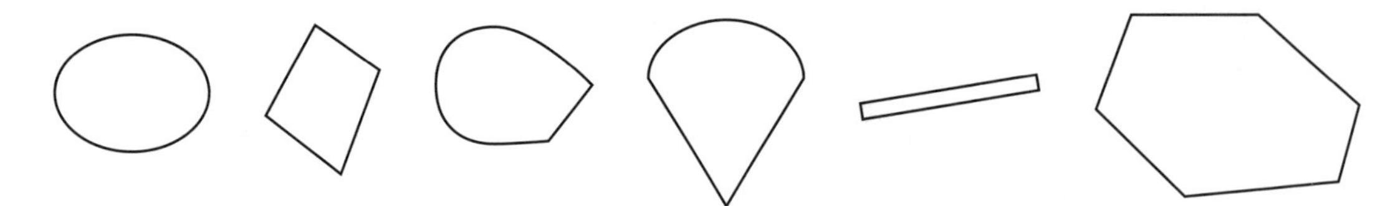

The figures below are *not* convex.

1. Circle the convex figures below.

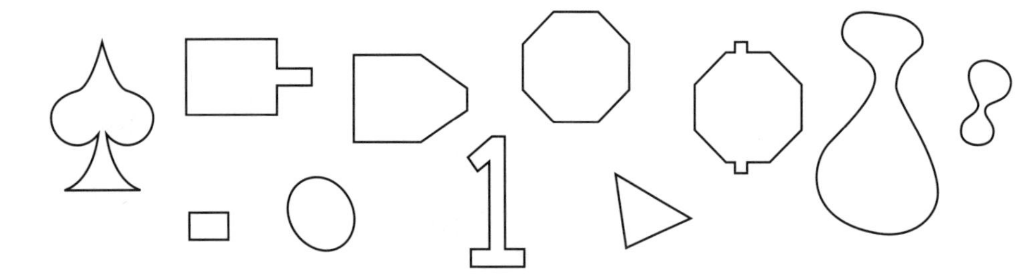

2. Draw a convex figure on the back of this sheet.
3. Draw a figure that is not convex on the back of this sheet.
4. Write down a definition of *convex* in your own words.

5. Find convex figures that can be made with tangram pieces. Sketch your solutions on the back of this sheet. (If you want an extra challenge, see how many seven-piece convex tangram figures you can find.)

Discussion

A. In your opinion, what makes a good tangram puzzle?

B. What makes a tangram puzzle easy or difficult?

3 Polygons

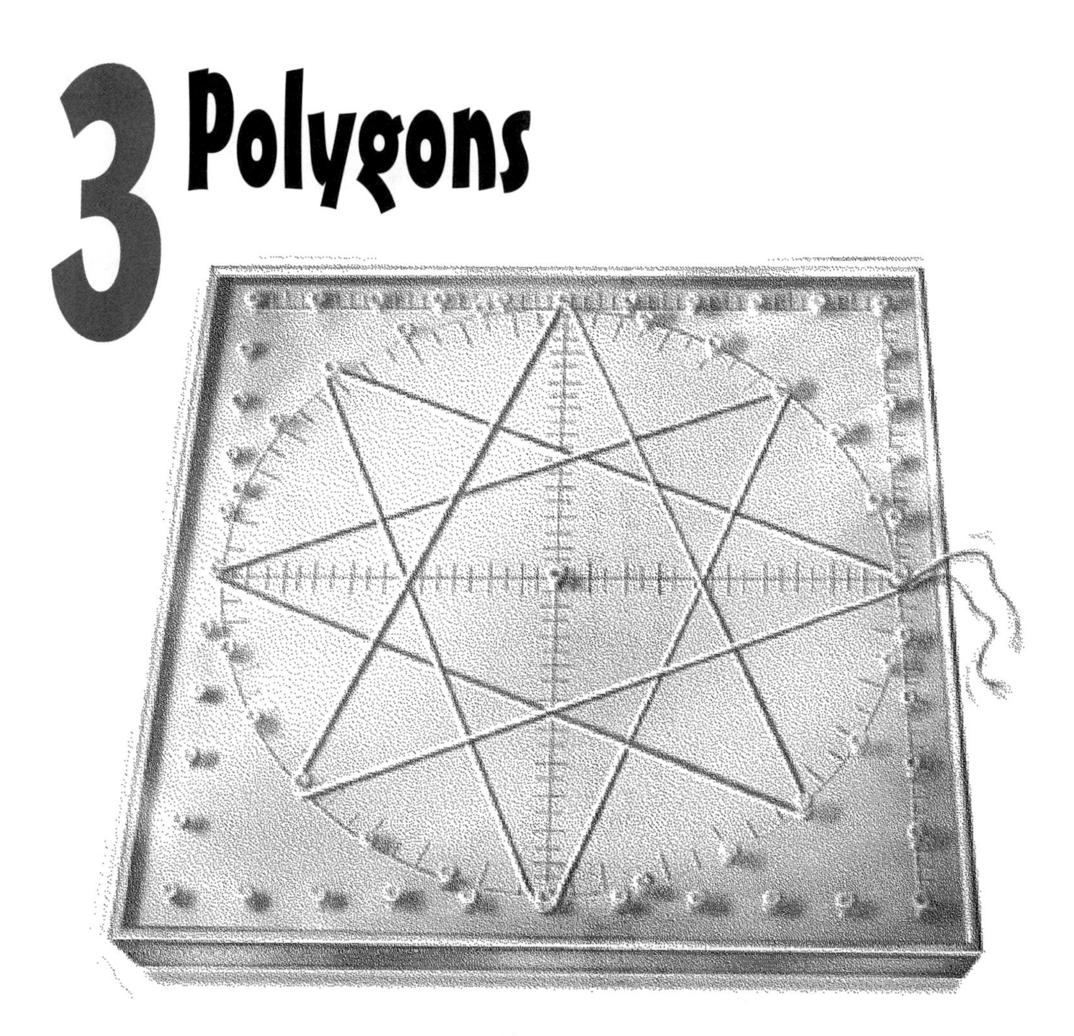

In the first two labs, the students construct triangles, given their sides and given their angles. In addition to the overt topics (the triangle inequality and the sum of the angles in a triangle), these labs allow you to introduce your chosen approach to construction (see the teacher notes to Lab 3.1). Moreover, they provide a first preview of two big ideas: congruence and similarity.

Labs 3.3, 3.5, and 3.6 ground students' understanding of angles by having them "walk around" the perimeters of various polygons. This complements the work from Section 1, and it also lays the groundwork for discovering the formula for the sum of the angles of a polygon. Turtle graphics, an approach to geometry that was initiated in the context of the computer language Logo, was the original inspiration for polygon walks. See the teacher notes pertaining to Lab 3.5 (Walking Regular Polygons) and Lab 3.6 (Walking Nonconvex Polygons).

The section ends with additional lessons on the sum of the angles of a polygon, which provide interesting algebraic connections.

See page 176 for teacher notes to this section.

LAB 3.1
Triangles from Sides

Name(s) ____________________

■ **Equipment:** Compass, straightedge, unlined paper

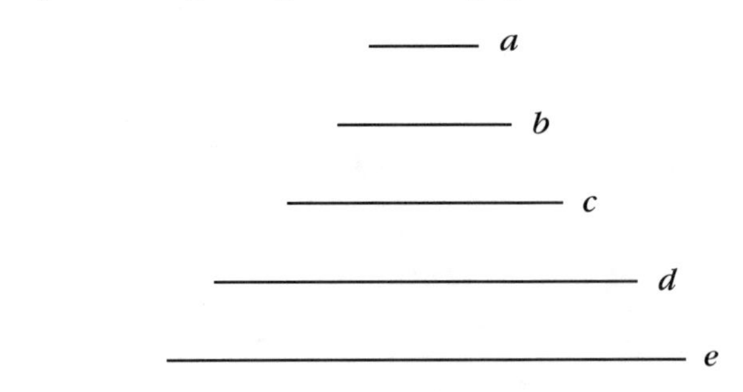

1. Here is a technique for constructing a triangle with sides equal to *c, d,* and *e* above. Practice this technique on a separate paper.
   - **a.** Copy side *e*. The endpoints give you two vertices of the triangle.
   - **b.** Make an arc centered at one end, with radius *d*.
   - **c.** Make an arc centered at the other end, with radius *c*.
   - **d.** Where the two arcs meet is the third vertex. Connect it to the other two. You are done.

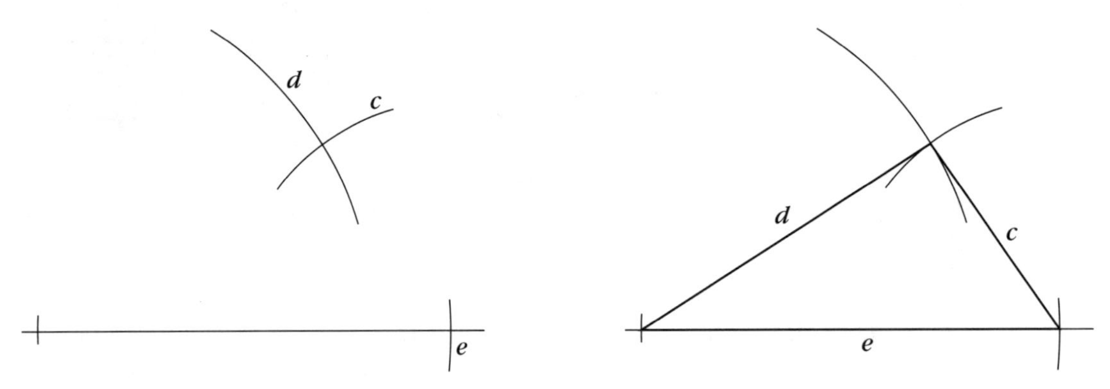

2. Using the same technique, construct at least three triangles with sides equal to *a, b, c, d,* or *e*. Try to make triangles different from those of your neighbors.

3. Try to construct a triangle with sides equal to *a, b,* and *e*. In the space below, explain why it is impossible.

4. Working with other students, make a complete list of possible triangles using sides *a, b, c, d,* and *e*. Include equilateral, isosceles, and scalene triangles. Also make a list of impossible triangles.

Possible (example: *c,d,e*) **Impossible** (example: *a,b,e*)

Discussion

A. Would it be possible to construct a triangle with side lengths 2, 4, and 8.5? Explain.

B. Would it be possible to construct a triangle with side lengths 2, 4, and 1.5? Explain.

C. If two of the sides of a triangle have lengths 2 and 4, what do you know about the third side?

D. State a generalization about the lengths of the three sides of a triangle.

LAB 3.2
Triangles from Angles

Name(s) ________________________

■ **Equipment:** Compass, straightedge, unlined paper

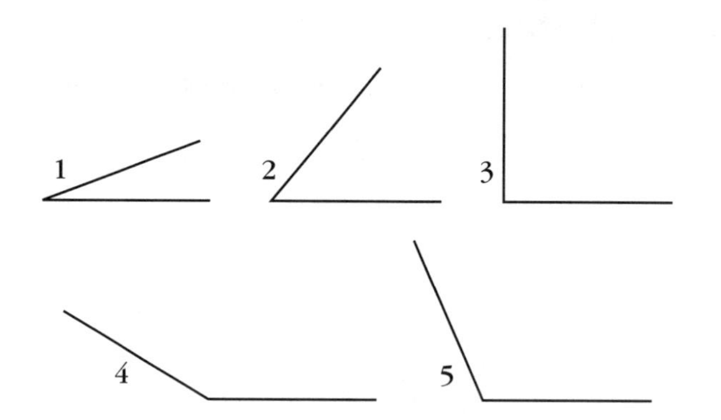

1. Here is a technique for constructing a triangle with angles equal to ∠1 and ∠2. Follow these steps and practice this technique on a separate sheet of paper.
   - a. Copy ∠1. That gives you one vertex.
   - b. Choose any point on a side of ∠1. That will be the second vertex.
   - c. Copy ∠2 at the second vertex. (Note that in the figure at right above, it was necessary to flip ∠2 over.)
   - d. Extend the other sides of ∠1 and ∠2. They meet at the third vertex. You are done.
2. Using the same technique, construct at least three triangles by copying two angles from among ∠1, ∠2, ∠3, ∠4, and ∠5. Try to make triangles different from those of your neighbors.
3. Would it be possible to construct a triangle with angles equal to ∠4 and ∠5? Explain in the space below.

4. Make a list of possible triangles using two angles chosen from among ∠1, ∠2, ∠3, ∠4, and ∠5 (including obtuse, right, and acute triangles). It's okay to use two copies of the same angle. Also make a list of impossible triangles.

Possible (example: ∠1, ∠2) **Impossible**

Discussion

A. Given three angles, is it usually possible to construct a triangle with those angles? What must be true of the three angles?

B. What must be true of two angles to make it possible to construct a triangle with those angles?

C. Given two angles, ∠1 and ∠2, describe two different ways to construct the angle $180° - (\angle 1 + \angle 2)$.

LAB 3.3 Walking Convex Polygons

Name(s) ______________________

■ **Equipment:** Pattern blocks, template, unlined paper

Definitions:

A *polygon* is a closed figure with straight sides.
An *n-gon* is a polygon with *n* sides.
A *convex* polygon is one where no angle is greater than 180°.

1. Is there such a thing as a nonconvex triangle? If so, sketch one in the space at right; if not, explain why not.
2. Is there such a thing as a nonconvex quadrilateral? Sketch one, or explain why there's no such thing.
3. For each number of sides from 3 to 12, make a *convex* pattern block polygon. Use your template to record your solutions on a separate sheet.

In this activity, you will give each other instructions for walking polygons. As an example, here are instructions for walking the pattern block trapezoid. We will use a scale of one step for one inch:

Walk forward one step from the starting point;
Turn right 60°;
Walk forward one step;
Turn right 60°;
Walk forward one step;
Turn right 120°;
Walk forward two steps;
Turn right 120°.

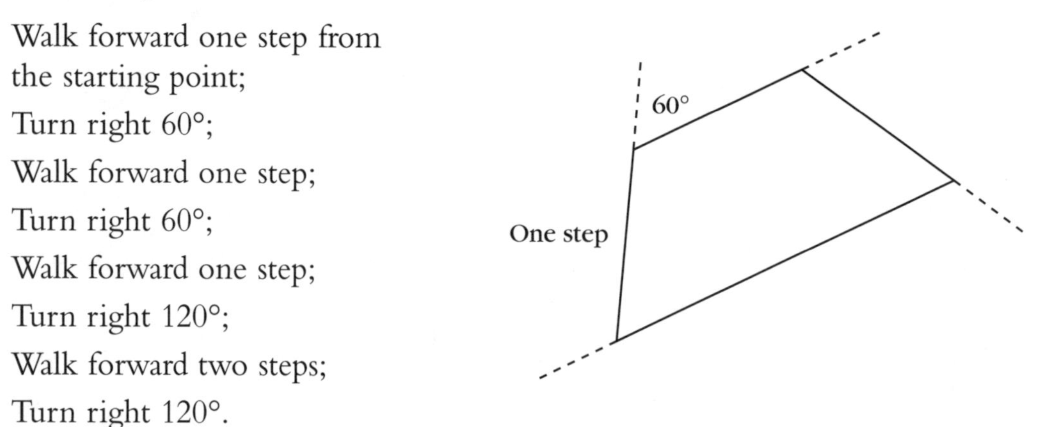

The final turn is not necessary, since the whole perimeter has been walked by then. However, it is traditional to include it, because it brings the walker back exactly to the starting position.

4. Mark the remaining steps and angles on the figure.

Note: The turn angle is also called the *exterior angle*. It is not usually the same as the interior angle, which is usually called simply the *angle*.

5. Write another set of instructions for walking the trapezoid, starting at another vertex.

6. On a separate paper, write instructions for walking each of the other pattern blocks.

7. Take turns with a partner, following each other's instructions for walking different blocks.

8. Make a two-block *convex* pattern block polygon, such as the one in the figure at right, and write instructions below for walking it.

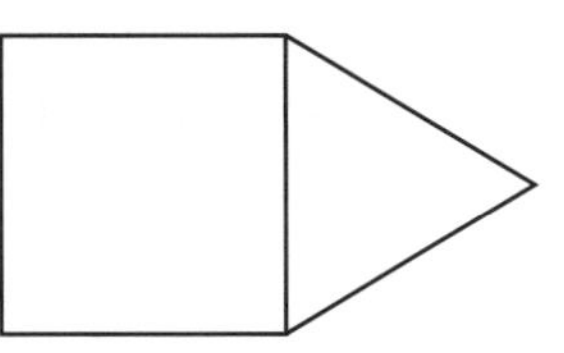

Discussion

A. Write alternate instructions for walking the trapezoid with one or more of these additional constraints:
   - Walking backward instead of forward
   - Making left turns instead of right turns
   - Starting in the middle of the long side

B. What is the relationship between the turn angle and the interior angle? When are they equal?

C. How can you tell whether a polygon walk is clockwise or counterclockwise by just reading the walk's instructions?

LAB 3.4

Name(s) ____________________

Regular Polygons and Stars

■ **Equipment:** Circle geoboard, string, Circle Geoboard Record Sheet

Definition: A *regular polygon* is a polygon in which all the angles are equal to each other and all the sides are equal to each other.

1. A regular triangle is called a(n) ______________ triangle. A regular quadrilateral is called a(n) ___________.

2. Tie a string at the 0° peg on the geoboard. Then make a figure by going around every fourth peg. The beginning of this process is shown in the figure at right.

   Keep going until you are back to the starting peg.

   a. What sort of figure did you get? ___________
   Sketch it on a circle geoboard record sheet.

   b. What is the measure of each angle of your figure?
   ____

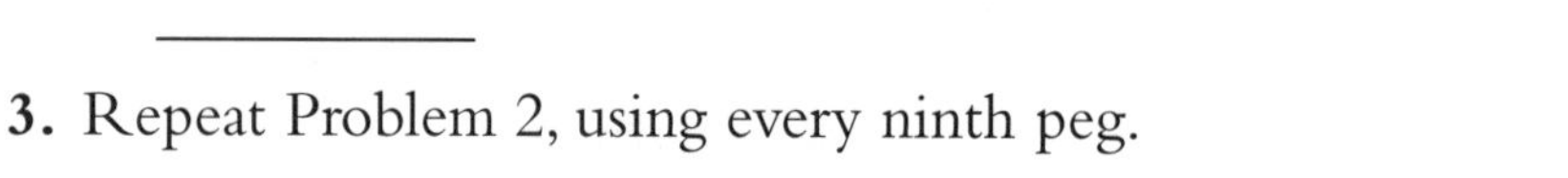

3. Repeat Problem 2, using every ninth peg.

Every *p*-th peg	Star or polygon?	Number of sides	Angle measure
1			
2			
3			
4	polygon	6	120°
5			
6			
7			
8			
9	star		
10			
11			
12			

Every *p*-th peg	Star or polygon?	Number of sides	Angle measure
13			
14			
15			
16			
17			
18			
19			
20			
21			
22			
23			
24			

LAB 3.4 Name(s) ______________________

Regular Polygons and Stars (continued)

4. Repeat Problem 2 by using every peg, every other peg, every third peg, and so on, all the way to every twenty-fourth peg. Look for patterns. You won't have enough string to make some of the stars on the geoboard, but you should still draw them on your record sheet. Record your entries in the table on the previous page, and record below any patterns you discover.

Discussion

A. Which of the pattern blocks are regular polygons?

B. Explain why the blue pattern block is not a regular polygon.

C. Make a pattern block polygon that has all angles equal but is not regular.

D. What patterns did you discover while filling out the table?

E. How can you predict whether using every p-th peg will give a star or a polygon?

F. Describe the star that you think would require the most string. Describe the star (not polygon) that would require the least string.

G. Discuss the situation when $p = 12$ or $p = 24$.

H. Explain why there are two values of p that yield each design.

I. How can you predict the angle from the value of p? Give a method or a formula.

J. Repeat this exploration on an imaginary 10- or 20-peg circle geoboard.

K. For a circle geoboard with n pegs, which p-th pegs will produce a star, and which will produce a polygon?

L. Investigate stars that use two strings or that require you to pick up your pencil, like the one shown at right.

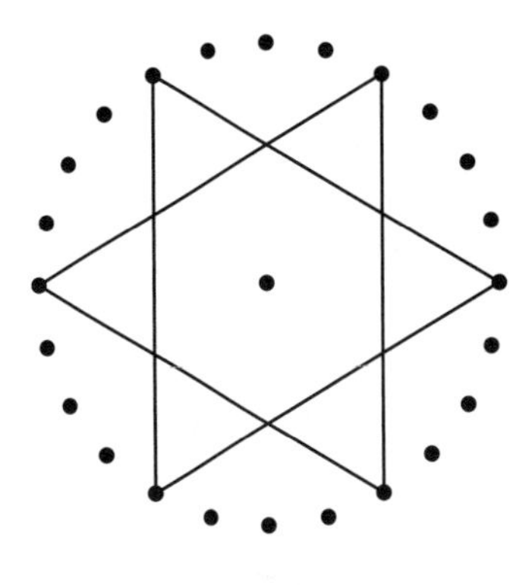

LAB 3.5

Name(s) ______________________

Walking Regular Polygons

Here are three descriptions of how to walk a square.

Jenny: Turn 90° left and take a step four times.
Maya: (Take step, turn right 90°) · 4
Pat: (1) Step forward; (2) turn right 90°; (3) step forward; (4) turn right 90°; (5) step forward; (6) turn right 90°; (7) step forward.

1. What is the *total turning* in degrees for each set of directions? (They are not all the same!) Jenny: __________ Maya: ______________ Pat: ___________

2. Use Maya's style to write instructions for an equilateral triangle walk.

3. What is the total turning for the triangle? _________

4. a. When we say that the sum of the angles in a triangle is 180°, what angles are we adding?

b. When we calculate the total turning, what angles are we adding?

5. These problems are about a regular 7-gon.

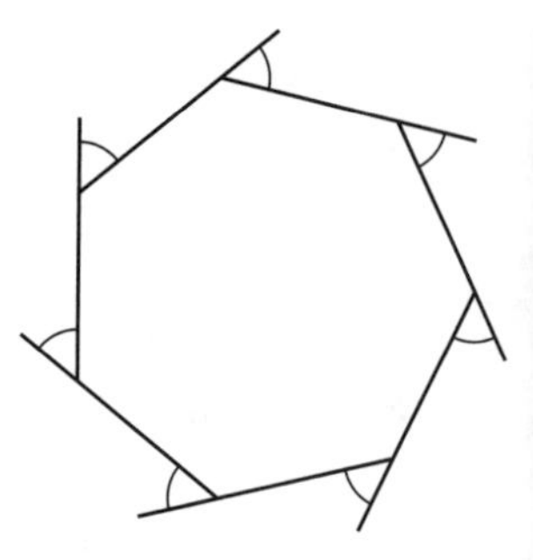

a. What is the measure of each exterior angle? Explain.

b. What is the measure of each interior angle? Explain.

6. Fill out this table for regular polygons.

Number of sides	Each angle	Angles' sum	Turn angle	Total turning
3	60°	180°		
4	90°	360°		
5				
6				
7				
8				
9				
10				
11				
12				
100				
n				

Discussion

A. Compare the three sets of instructions for walking a square.

- **a.** Do they all work?
- **b.** How are they the same?
- **c.** How are they different?
- **d.** Which do you think is the clearest?
- **e.** How would you improve each of them?

B. Which column of the table in Problem 6 is easiest to find? How can this help you find the numbers in the other columns?

C. Compare the formulas you found with those of other students.

D. Explore the angles for regular stars: the sum of interior angles; the sum of turn angles. Can you find patterns? Formulas?

LAB 3.6

Name(s) ______________________

Walking Nonconvex Polygons

■ **Equipment:** Pattern blocks

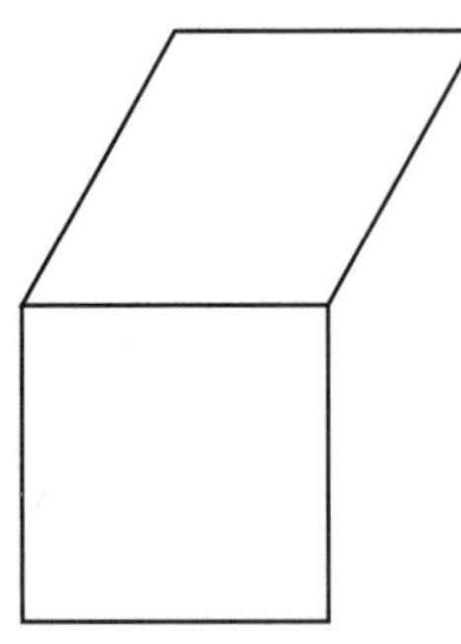

1. Use two pattern blocks to create a nonconvex polygon such as the one in the figure at right.

   a. Write instructions for walking it, assuming that the walker begins and ends facing the same direction.

   b. Follow a neighbor's instructions, and have him or her follow yours.

2. Since the walker ends up facing the way he or she started, what should the *total turning* be?

   a. Calculate the total turning to check whether your prediction worked. Show your calculation below.

   b. Check a neighbor's total turning.

The direction you are facing is called your *heading*. North is heading 0°. Other headings are measured in degrees clockwise from North. East is 90°.

3. What is your heading if you are facing the following directions?

   a. South __________

   b. West __________

   c. Southwest __________

   d. NNW __________

4. Simone and Joy are standing back to back. Find Simone's heading if Joy's is:

   a. 12° __________

   b. 123° __________

   c. h° (careful!) __________

5. How would you interpret a heading greater than 360°?

6. How would you interpret a heading less than 0°?

7. Assume you start walking your nonconvex polygon facing North. What is your heading while walking each side?

8. Does turning left add to or subtract from your heading? What about turning right?

9. In view of Problem 8, how should you deal with turn angles to get results that are consistent with changes in heading? With this interpretation, what is the total turning for a polygon, whether convex or not? Explain.

Discussion

A. In Problem 2, the total turning should be 360°. How could we calculate total turning so that this works? (**Hint:** The solution involves using positive and negative numbers.)

B. What does turning 360° do to your heading? What about turning 350°? How is this related to the discussion about positive and negative turns?

C. What does it mean to turn right –90°? Can the same turn be accomplished with a positive left turn? A positive right turn?

D. If we want total turning to be 360° and want our turns to be consistent with changes in heading, should we walk polygons in a clockwise or counterclockwise direction?

LAB 3.7
Diagonals

Name(s) ____________________

Definition: A *diagonal* is a line segment connecting two *nonconsecutive* vertices of a polygon.

The figure at right is a regular decagon, with all its diagonals. How many diagonals are there? It is not easy to tell. In this activity, you will look for a pattern in the number of diagonals in a polygon. You will also use diagonals to think about the sum of the angles in a polygon.

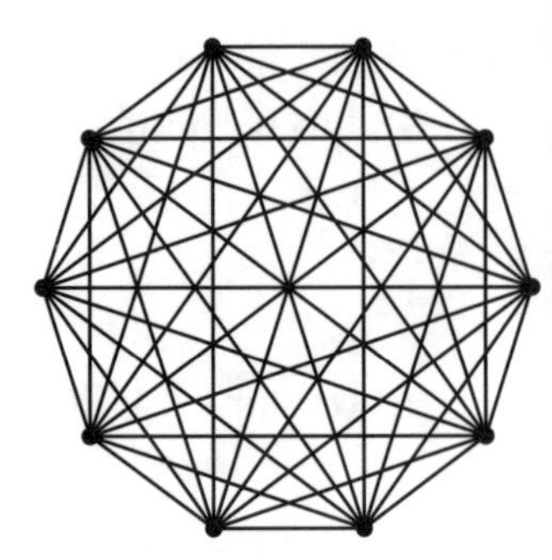

1. How many diagonals does a triangle have? __________

2. How many diagonals does a quadrilateral have? __________

3. Fill out this table, stating the total number of diagonals.

Sides	Diagonals
3	
4	
5	
6	

Sides	Diagonals
7	
8	
9	
100	

4. Find a formula for the total number of diagonals in an n-gon. ____________

In the remainder of the lab, we will limit our discussion to convex polygons, so all diagonals will be on the inside of the polygon.

5. If you draw all the diagonals *out of one vertex* in a convex n-gon, you have divided it into triangles—how many? (Experiment with 3-, 4-, 5-gons, and so on, and generalize.) __________

6. Use the answer to the previous problem to find a formula for the sum of the angles in a convex n-gon. __________

7. Explain why this reasoning does not work for a nonconvex n-gon.

Discussion

A. What patterns do you see in the table in Problem 3?

B. Experiment with convex and nonconvex polygons and their diagonals. Write a new definition of *convex polygon* that uses the word *diagonals*.

LAB 3.8

Name(s) ______________________

Sum of the Angles in a Polygon

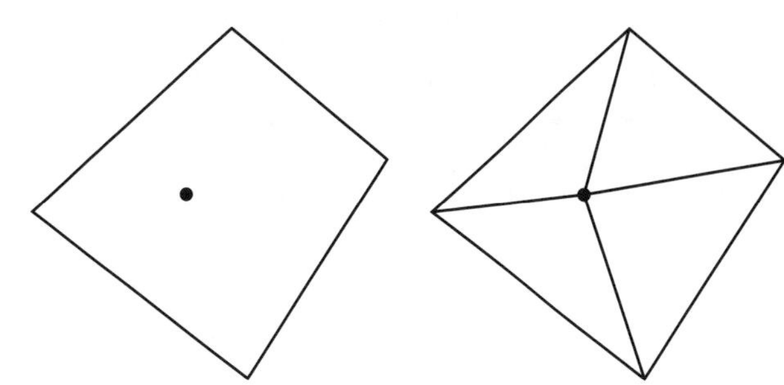

1. The sum of the interior angles of *any* quadrilateral is always the same. Here is an explanation of why that's true, based on dividing the quadrilateral into triangles.

a. Draw any random quadrilateral (not a square).

b. Put a point inside it, and connect it to each vertex.

c. What is the sum of all the angles in the four triangles? __________

d. How much of that sum is around the vertex at the center? __________

e. How much of that sum is the sum of the angles in the quadrilateral? __________

2. a. Use the same logic to find the sum of the angles in a 5-, 6-, 8-, and 12-gon.

5-gon __________

6-gon __________

8-gon __________

12-gon __________

b. Use the same logic to find a formula for the sum of the interior angles in an *n*-gon.

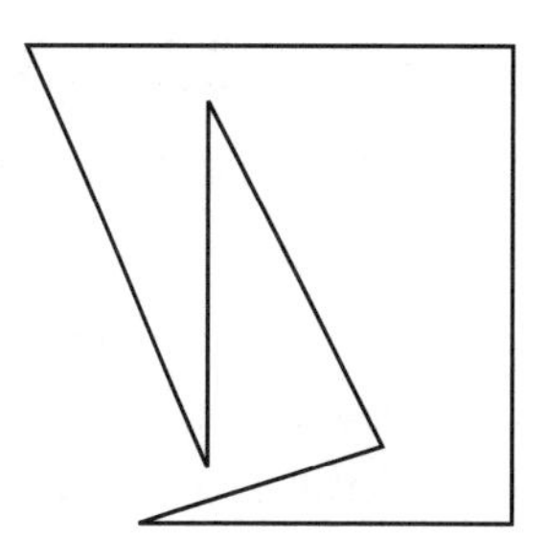

3. Use the figure at right, or one you create, to explain why the logic of Problems 1 and 2 does not work for all polygons.

4. Find a way to divide the polygon in Problem 3 into triangles in order to find the sum of its angles. Record this sum. ______________

5. Is your answer to Problem 4 consistent with the formula you found in Problem 2?

Discussion

A. Compare the method in Problem 1 for finding the sum of the angles in a polygon with the one in Lab 3.7. Do they yield the same formula? Do they use the same numbers of triangles?

LAB 3.9

Name(s) ____________________

Triangulating Polygons

Definition: To *triangulate* a polygon means to divide it into triangles. Triangle vertices can be vertices of the original polygon, or they can be new points, either on the inside of the polygon or on one of the sides.

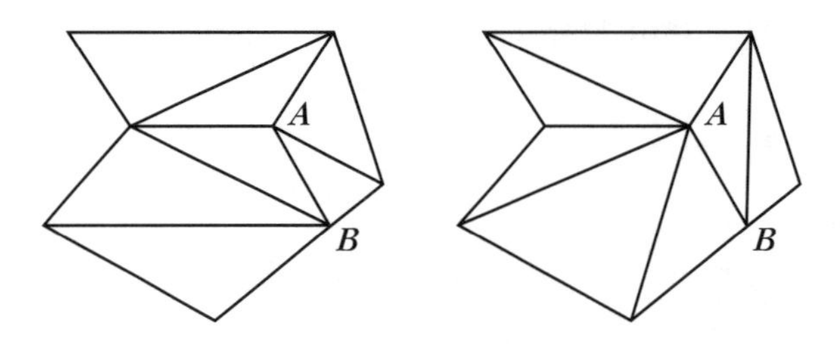

For example, the hexagon above has been divided into seven triangles in two different ways, both of them using one inside vertex (A) and one side vertex (B).

To triangulate a polygon, decide how many additional vertices you will use and where you will put them. Then connect vertices in any way you want, as long as the segments you draw never cross each other, and the polygon ends up divided into triangles.

Experiment with triangulating polygons. For each triangulation, enter the information in a table like the one below. Look for patterns. On a separate paper, write several paragraphs describing your findings. These questions may help you organize your research:

- How many triangles are added when you add 1 to the number of sides and keep other things constant?
- How many triangles are added when you add 1 to the number of inside vertices?
- How many triangles are added when you add 1 to the number of side vertices?
- Can you find a formula relating the number of original vertices (n), the number of inside vertices (i), the number of side vertices (s), and the number of triangles (t)?

The data for the triangulations above have been entered for you.

Original vertices	Inside vertices	Side vertices	Number of triangles
6	1	1	7

Name(s) ______________________

Discussion

A. In the example given, is it possible to get a different number of triangles by moving *A* to another location inside the hexagon and/or moving *B* to another location on one of the sides?

B. How would you define a minimal triangulation for a polygon? How might you draw it?

C. If the vertices are distributed differently, how does that affect the total number of triangles? For example, if you have a total of eight vertices, as in the example on the previous page, but with four original vertices, two inside vertices, and two side vertices, do you still have seven triangles? If not, how would you distribute the vertices to get the smallest number of triangles possible? What is that number? How would you distribute the vertices to get the greatest number of triangles possible? What is that number?

D. How is triangulation related to the sum of the angles in a polygon?

4 Polyominoes

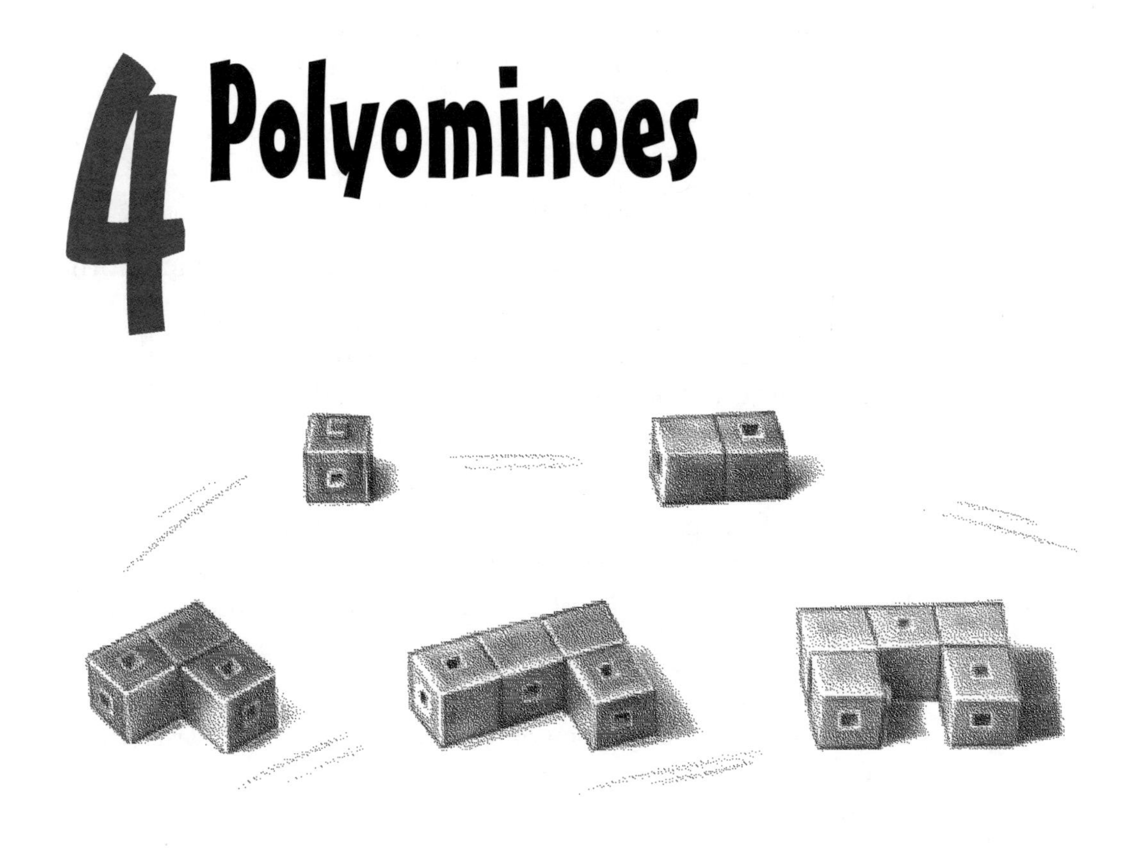

Polyominoes are a major topic in recreational mathematics and in the field of geometric combinatorics. Mathematician Solomon W. Golomb named and first studied them in 1953. He published a book about them in 1965, with a revised edition in 1994 *(Polyominoes: Puzzles, Patterns, Problems, and Packings)*. Martin Gardner's "Mathematical Games" column in *Scientific American* popularized many polyomino puzzles and problems. Since then, polyominoes have become one of the most popular branches of recreational mathematics. Mathematicians have created and solved hundreds of polyomino problems and have proved others to be unsolvable. Computer programmers have used computers to solve some of the tougher puzzles.

Polyominoes have connections with various themes in geometry—symmetry, tiling, perimeter, and area—that we will return to in future sections. The only part of this section that is required in order to pursue those connections is the introductory lab, Lab 4.1 (Finding the Polyominoes). The remaining labs are not directly related to topics in the traditional curriculum, but they help develop students' visual sense and mathematical habits of mind, particularly regarding such skills as:

- systematic searching;
- classification; and
- construction of a convincing argument.

These habits are the main payoff of these labs, as specific information about polyominoes is not important to students' further studies.

Use this section as a resource for students to do individual or group projects, or just as an interesting area for mathematical thinking if your students are enthusiastic about polyominoes.

In the world of video games, tetrominoes are well known as the elements of the game *Tetris,* which is probably familiar to many of your students. In fact, the figures in that game are *one-sided* tetrominoes (see Discussion Question C in Lab 4.2 (Polyominoes and Symmetry)).

Pentominoes (polyominoes of area 5) have enjoyed the greatest success among recreational mathematicians, gamers, and puzzle buffs. A commercial version of a pentomino puzzle called *Hexed* can often be found in toy stores. My book *Pentomino Activities, Lessons, and Puzzles,* available from Creative Publications along with plastic pentominoes, helped bring pentominoes into the classroom.

As a sequel to this work, see Lab 8.1 (Polyomino Perimeter and Area).

See page 184 for teacher notes to this section.

LAB 4.1
Finding the Polyominoes

Name(s) ______________________

■ **Equipment:** 1-Centimeter Grid Paper, interlocking cubes

This is a *domino*. It is made of two squares, joined edge to edge.

A *tromino* is made of three squares. This one is called the straight tromino.

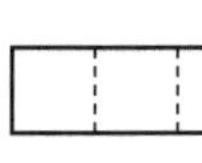

This is the *bent* tromino.

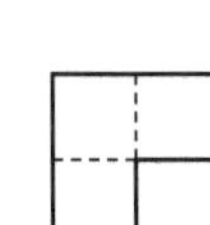

There are only two different trominoes. These are the same ones as above, but in different positions.

However, this is not a tromino, since its squares are not joined edge to edge.

Definition: Shapes that are made of squares joined edge to edge are called *polyominoes*.

You can make polyominoes using interlocking cubes. Be sure that when the figure is laid flat, all the cubes touch the table.

1. *Tetrominoes* are made of four squares. Find them all and record them on grid paper.
2. *Pentominoes* are made of five squares. Find them all and record them on grid paper.

Discussion

A. Find a way to convince an interested person that you have indeed found all of the polyominoes with area from 1 to 5, and that you have no duplicates.

B. A natural way to classify the polyominoes is by area. Find other ways to classify them.

C. Which pentominoes can be folded into a box without a top?

Polyomino Names Reference Sheet

These are the standard polyomino names.

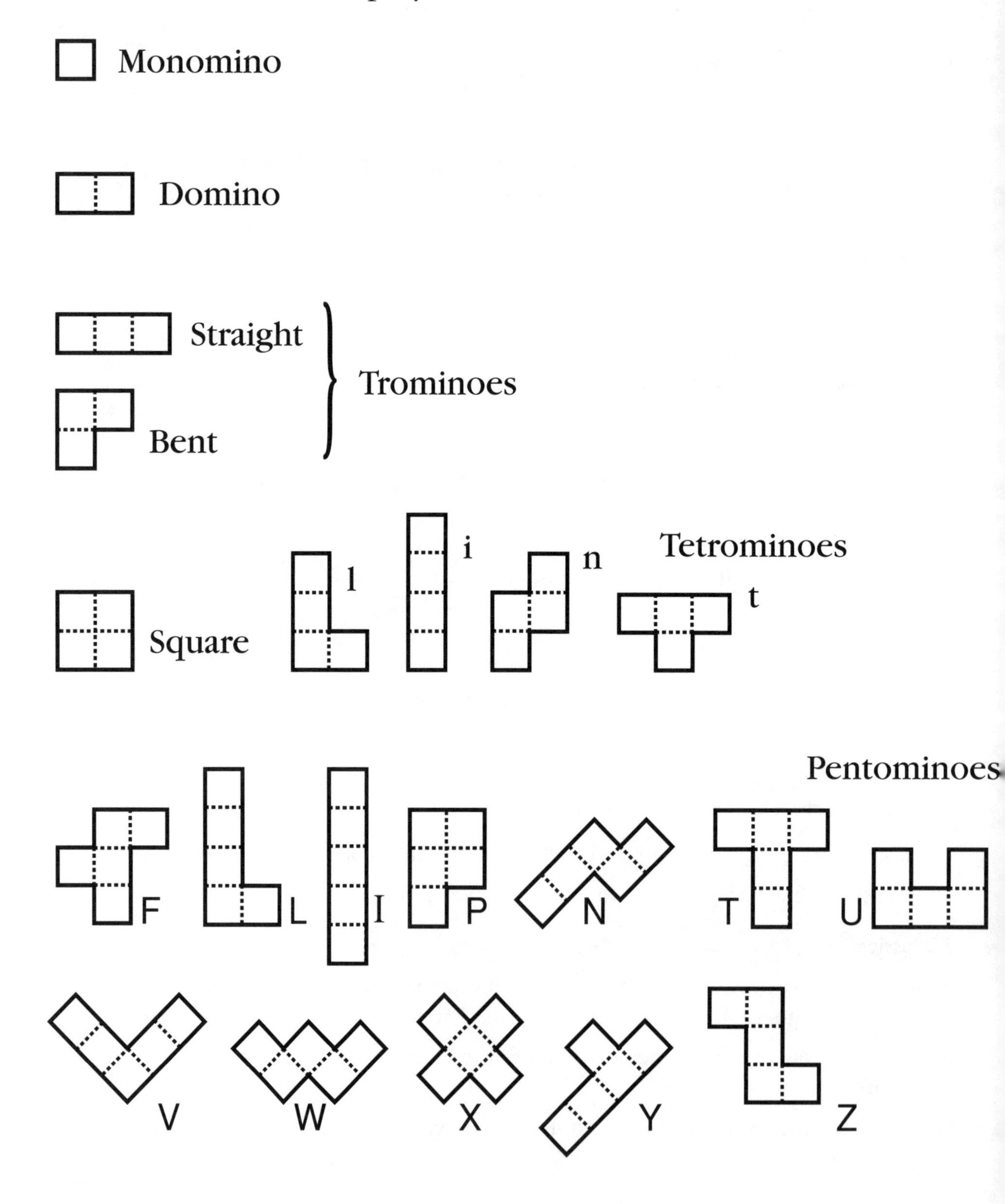

LAB 4.2

Name(s) ______________________

Polyominoes and Symmetry

- **Equipment:** 1-Centimeter Grid Paper, interlocking cubes, Polyomino Names Reference Sheet

The straight tromino can be placed on graph paper in two different ways:

1. For each of the polyominoes with area from 1 to 5, how many different positions are there on graph paper?

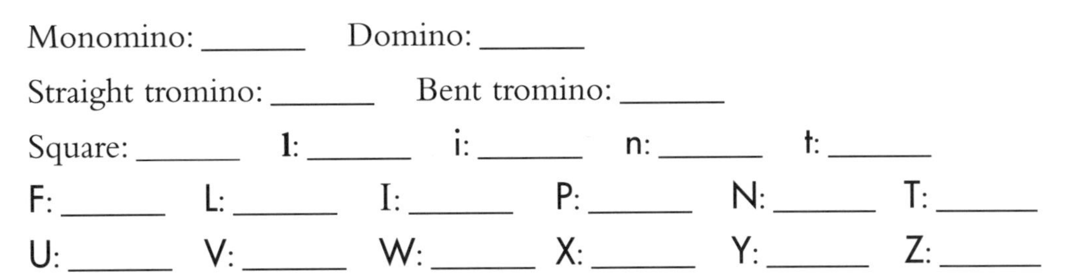

Monomino: ______ Domino: ______

Straight tromino: ______ Bent tromino: ______

Square: ______ l: ______ i: ______ n: ______ t: ______

F: ______ L: ______ I: ______ P: ______ N: ______ T: ______

U: ______ V: ______ W: ______ X: ______ Y: ______ Z: ______

Trace the straight tromino. You can rotate it 180° and it will still fit on its outline. We say that it has two-fold *turn symmetry* or *rotation symmetry*. Alternatively, you can flip it around a vertical axis, or a horizontal one, and it will still fit. We say that it has two *lines of symmetry,* or *mirror lines.*

2. Write the name of each polyomino with area 1 to 5 in the appropriate space in the table below. Some spaces may be empty; others may have more than one entry. For example, the straight tromino is *not* the only polyomino with two mirror lines and two-fold turn symmetry.

	No mirror lines	**1 mirror line**	**2 mirror lines**	**3 mirror lines**	**4 mirror lines**
No turns (except 360°)					
Two-fold turn (180°)			straight tromino		
Three-fold turn (120°)					
Four-fold turn (90°)					

Name(s) ______________________

Discussion

A. Which types of symmetry in Problem 2 have no polyomino examples? Why?

B. How are the answers to Problem 1 and Problem 2 related?

C. Imagine that polyominoes were one-sided (like tiles) and that you could not flip them over. Then there would have to be, for example, separate pentominoes for the shapes L and J. In other words, there would be more polyominoes. Find all the additional polyominoes you would need in order to have a full set for areas 1 to 5.

LAB 4.3

Name(s) ______________________

Polyomino Puzzles

■ **Equipment:** 1-Centimeter Grid Paper, interlocking cubes

1. Using interlocking cubes, make a set of all of the polyominoes with area 2, 3, and 4. You should have eight pieces. It is best to make each polyomino a single color.
2. On graph paper, draw all the rectangles (including squares) that satisfy these two conditions.
   - They have area 28 or less.
   - Their dimensions are whole numbers greater than 1.
3. Use the polyominoes you made in Problem 1 to cover the rectangles you drew in Problem 2. The 2 × 3 rectangle, covered by a domino and 1 tetromino, is shown in the figure at right. Record your solutions. **Note:** One rectangle is impossible with this set of polyominoes.
4. On Centimeter Grid Paper, draw staircases like the ones in the figures at right, with area between 3 and 28, inclusive. Cover each one with some of the polyominoes you made in Problem 1, and record your solutions.

Discussion

A. Describe your system for finding all the rectangles in Problem 2.

B. Which rectangle puzzle *cannot* be solved in Problem 3? Why?

C. Find a formula for the area of a staircase like the two-step and three-step examples in Problem 4, given that it has x steps.

LAB 4.4
Family Trees

Name(s) ____________________

■ **Equipment:** 1-Centimeter Grid Paper, interlocking cubes

A polyomino is a *child* of another polyomino if it comes from the original polyomino by the addition of a single square. For example, the **l**, **i**, and **t** tetrominoes are children of the straight tromino. The square and **n** tetrominoes are not.

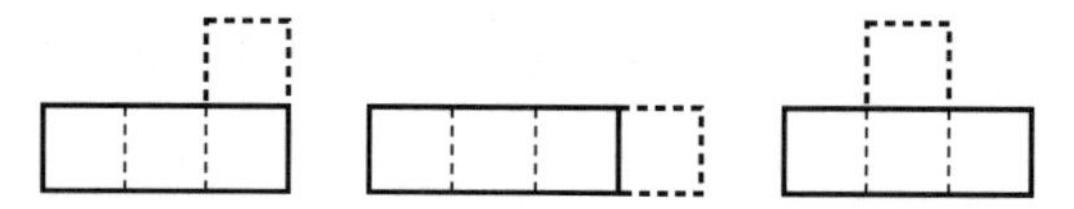

Here is a family tree for the **t** tetromino. It shows all of its ancestors back to the monomino, and all of its pentomino children.

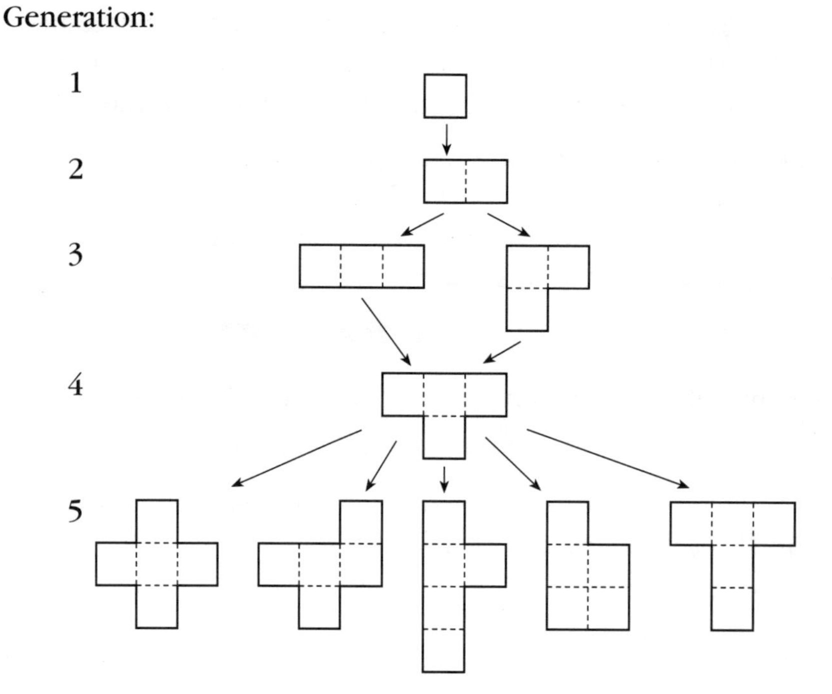

On grid paper, make complete family trees for the remaining tetrominoes.

1. l

2. Square

3. i

4. n

Discussion

A. Which pentomino has the most (tetromino) parents?

Two polyominoes "of the same generation" are called siblings (brothers and sisters) if they have a parent in common. For example, the F and the P pentominoes are siblings; both are children of the t tetromino.

B. List all the siblings of the I pentomino.

C. List all the siblings of the W pentomino.

Polyominoes of the same generation are cousins if they are *not* siblings. For example, the square and the straight tetromino are cousins because they do not have a parent in common.

D. List all the cousins of the Y pentomino.

E. Find two "second cousin" pentominoes. (These are pentominoes that have no tetromino or tromino ancestors in common.)

F. Which pentomino has the most (hexomino) children?

G. Which pentomino has the fewest (hexomino) children?

LAB 4.5
Envelopes

Name(s) ____________________

- **Equipment:** 1-Centimeter Grid Paper, interlocking cubes

What is the smallest rectangle (or square) onto which you can fit the bent tromino? As you can see in this figure, it is a 2×2 square.

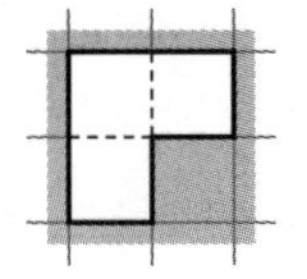

A 1×3 rectangle is the smallest rectangle onto which you can fit the straight tromino. We say that the 1×3 and 2×2 rectangles are tromino *envelopes*.

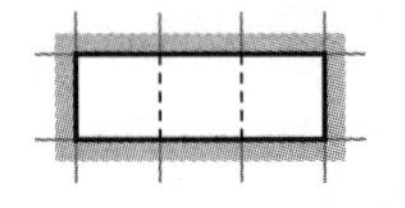

The 2×3 rectangle is not a tromino envelope, because it is too big.

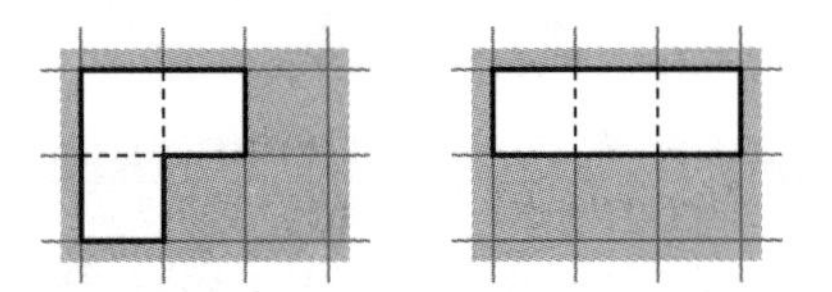

1. There are three tetromino envelopes. Find them and sketch them on grid paper. Which tetrominoes belong with each?
2. There are four pentomino envelopes. Find them and sketch them on grid paper. Which pentominoes belong with each?

The 3×6 rectangle is too big to be a hexomino envelope, as shown here.

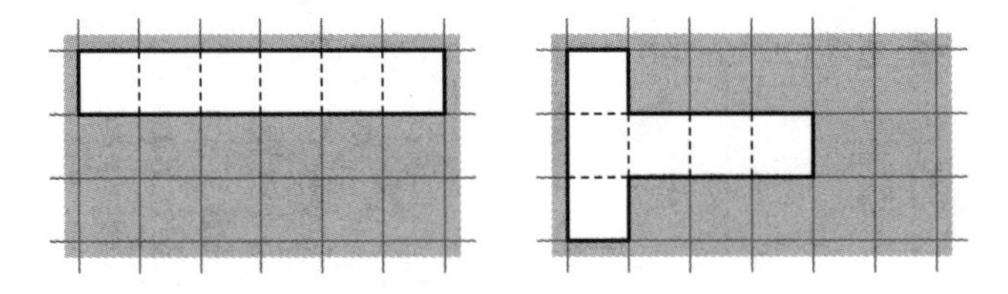

The 3×3 square, however, is a hexomino envelope.

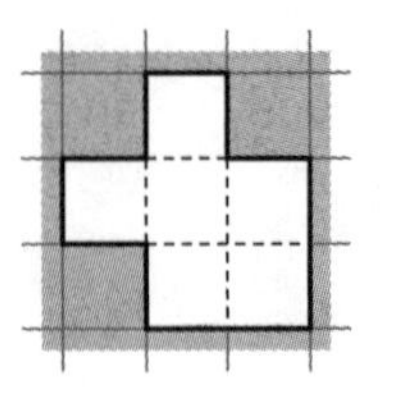

Name(s) ______________________

3. Here is a list of rectangles.

2×2	1×5	2×3	2×5
3×3	3×5	2×4	1×6
3×4	2×6	4×4	3×6

Experiment on grid paper, then perform the following three steps.

a. Cross out those rectangles in the list above that are too small to be hexomino envelopes.

b. Cross out those rectangles in the list above that are too big to be hexomino envelopes.

c. Make sure the remaining rectangles in the list are hexomino envelopes!

4. Find as many hexominoes as you can. Sketch them on grid paper.

Discussion

A. In some cases, the perimeter of the envelope is the same as the perimeter of the corresponding polyomino. Is it ever true that the perimeter of the envelope is greater than that of the polyomino? Is it ever true that it is less? Explain.

B. Which hexominoes can be folded into a cube?

LAB 4.6
Classifying the Hexominoes

Name(s) ______________________

■ **Equipment:** 1-Centimeter Grid Paper, interlocking cubes

There are 35 hexominoes. Is that the number you found in Lab 4.5? Organize them by drawing them in the envelopes below. As you work, watch out for duplicates. There should be no empty envelopes when you finish.

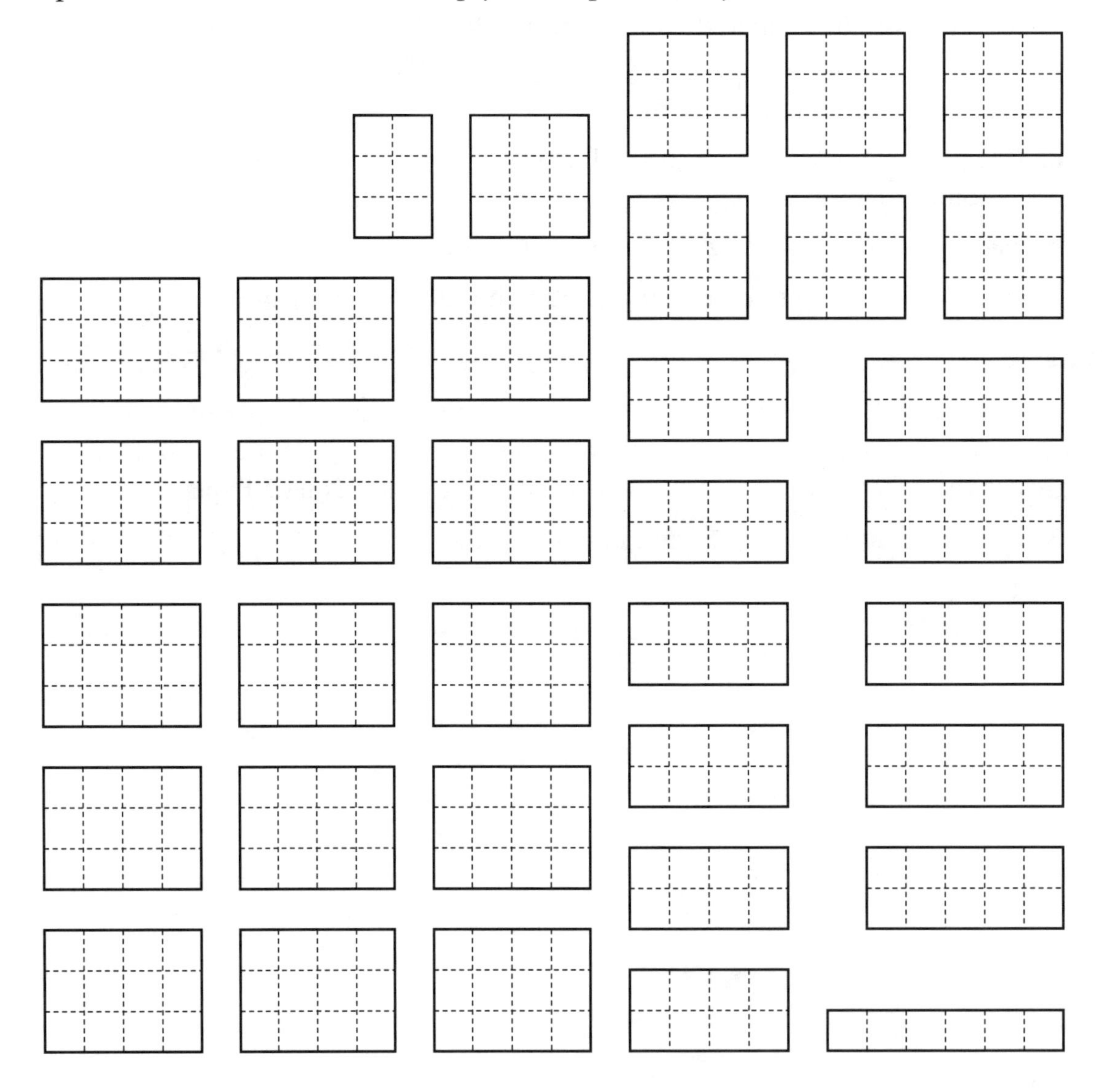

LAB 4.7
Minimum Covers

Name(s) ______________________

■ **Equipment:** 1-Centimeter Grid Paper, interlocking cubes

The squares covered by shading below form the smallest shape onto which you can fit either tromino. Its area is 4 square units. This shape is called a *minimum tromino cover.*

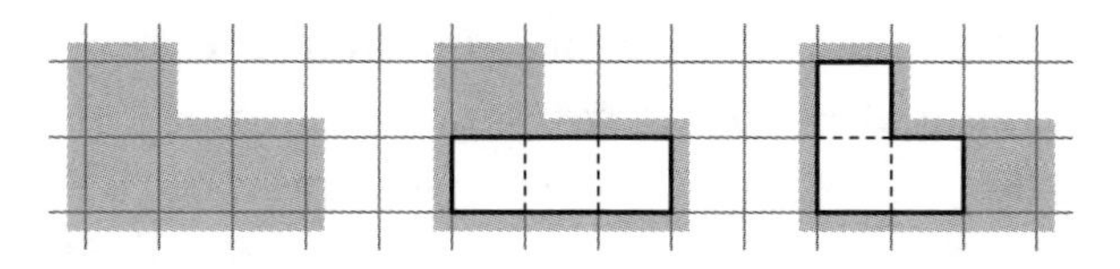

All of the tetrominoes fit on the shape covered by shading below, which has an area of 7 square units. It is not, however, the minimum tetromino cover.

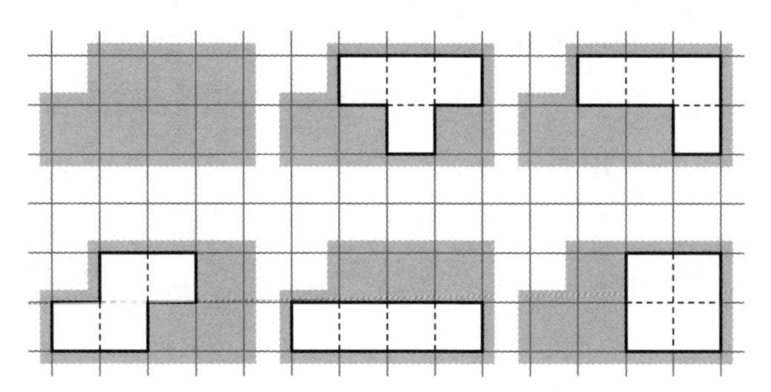

1. Find the smallest shape onto which any of the tetrominoes will fit (the minimum tetromino cover).
   a. Sketch it on grid paper.
   b. What is its area?
2. Find the minimum pentomino cover.
   a. Sketch it on grid paper.
   b. What is its area?
3. Find the minimum hexomino cover.
   a. Sketch it on grid paper.
   b. What is its area?

Discussion

A. How did you find the minimum cover? How do you know that no smaller cover is possible?

B. Is the minimum cover unique? In other words, can more than one shape be used for a minimum cover?

LAB 4.8
Polycubes

Name(s) ______________________

■ **Equipment:** Interlocking cubes

Polycubes are three-dimensional versions of polyominoes. They are made by joining cubes face to face. At right is an example (a pentacube).

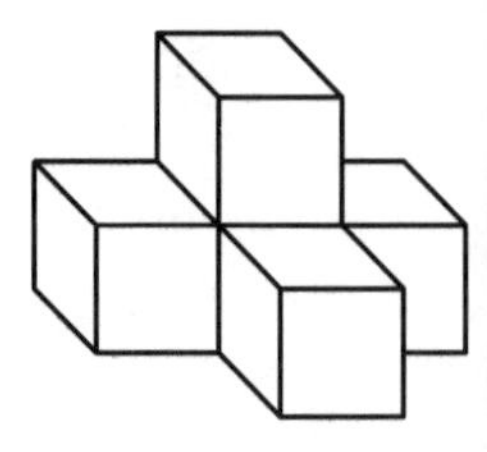

1. Use your cubes to make one monocube, one dicube, two tricubes, and eight tetracubes. **Hint:** Most of them look just like the corresponding polyominoes, except for three of the tetracubes, which are genuinely 3-D. Two of them are mirror images of each other.
2. Put aside all the polycubes that are box-shaped. The remaining pieces should have a total volume of 27. They can be combined to make the classic Soma® Cube.
3. Find other interesting figures that can be made using the Soma pieces.
4. Find all the pentacubes. Twelve look like the pentominoes, and the rest are genuinely 3-D.
5. What rectangular boxes can you make using different pentacubes?

Discussion

A. How did you organize your search for polycubes? (Did you use the polyominoes as a starting point? If yes, how? If not, what was your system?) How did you avoid duplicates?

B. What are some ways to classify the polycubes?

C. What would be the three-dimensional equivalent of a polyomino envelope?

D. Make a family tree for a tetracube.

LAB 4.9
Polytans

Name(s) ______________________

■ **Equipment:** Graph paper

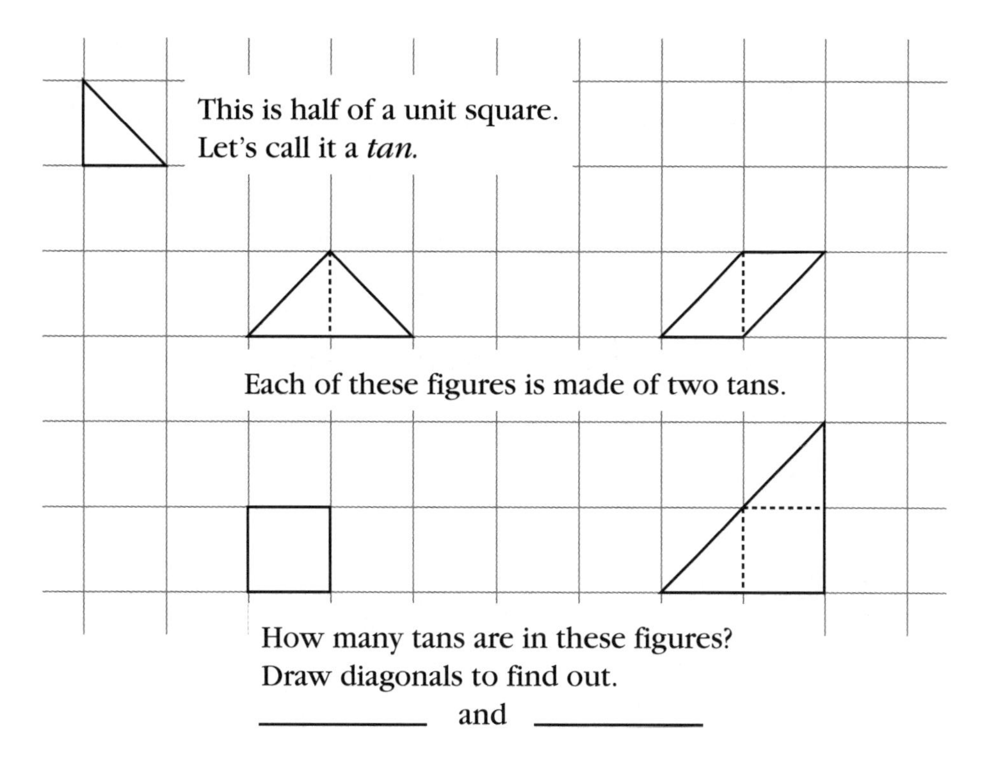

This is half of a unit square.
Let's call it a *tan*.
Each of these figures is made of two tans.
How many tans are in these figures?
Draw diagonals to find out.
__________ and __________

1. There are only four figures that are made of three tans. Use graph paper to find them.

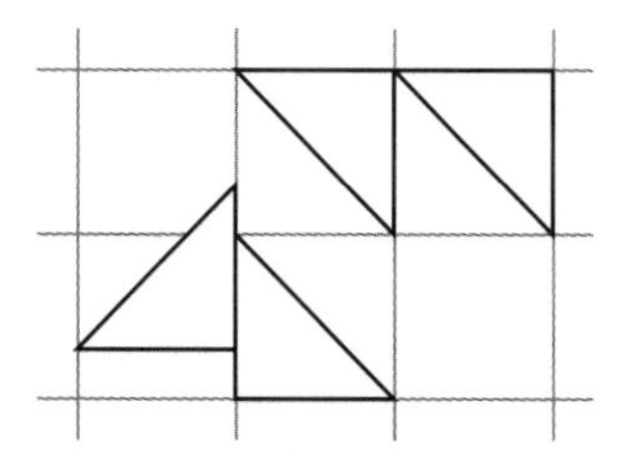

Note: Arrangements like these do not count. Tans must touch by whole sides.

2. How many figures can you find that are made up of four tans? Draw them on graph paper.

Discussion

A. What is your method for finding 4-tan shapes?

B. How do you know when you have found all the 4-tan shapes?

LAB 4.10
Polyrectangles

Name(s) ______________________

■ **Equipment:** Template, graph paper

Polyrectangles are like polyominoes, except that their basic unit is a rectangle that is not a square. The rectangle edges that meet to form a polyrectangle must be equal. There is only one domino, but there are two directangles.

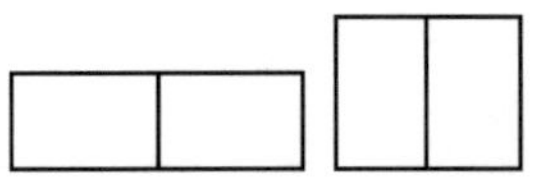

The two directangles

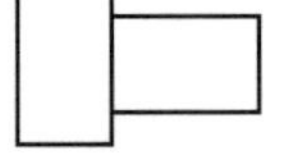

Not a directangle

Record your findings for the following questions on a separate sheet of paper.

1. Find all the trirectangles.
2. Find all the tetrarectangles.
3. Find all the pentarectangles.

If we add the constraint that polyrectangles cannot be turned over, additional figures are needed. For example, there are four one-sided trirectangles, shown below. The second and fourth examples would be duplicate trirectangles, but as one-sided trirectangles they are different.

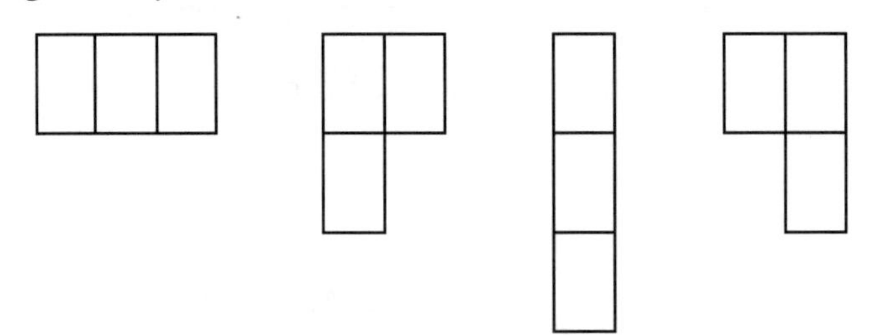

4. Find all the one-sided tetrarectangles.

Yet another variation is to think of polyrectangles as polystamps, torn out of a sheet of stamps, with each rectangle displaying a figure and some text. You cannot turn polystamps at all. Up is up. In that situation, we have six tristamps:

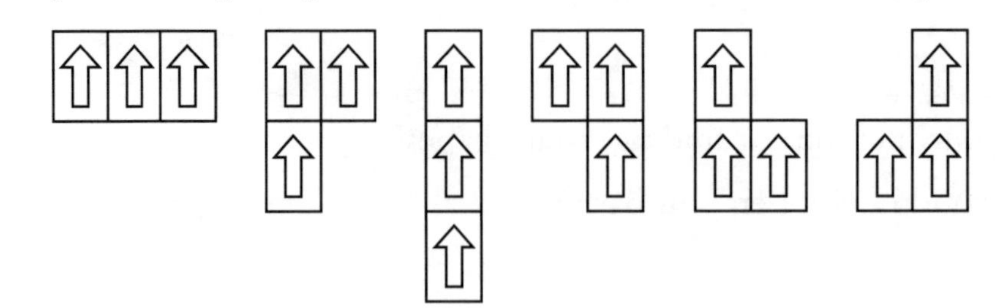

5. Find all the tetrastamps.

LAB 4.10

Name(s) ______________________

Polyrectangles (continued)

Discussion

A. Each polyrectangle corresponds to one polyomino. However, a polyomino may correspond to one or two polyrectangles. What determines whether it is one or two?

B. What is the relationship between polystamps and polyrectangles?

C. The more symmetries a polyomino has, the more polyrectangles, one-sided polyrectangles, and polystamps there are that are related to it. True or false? Explain.

5 Symmetry

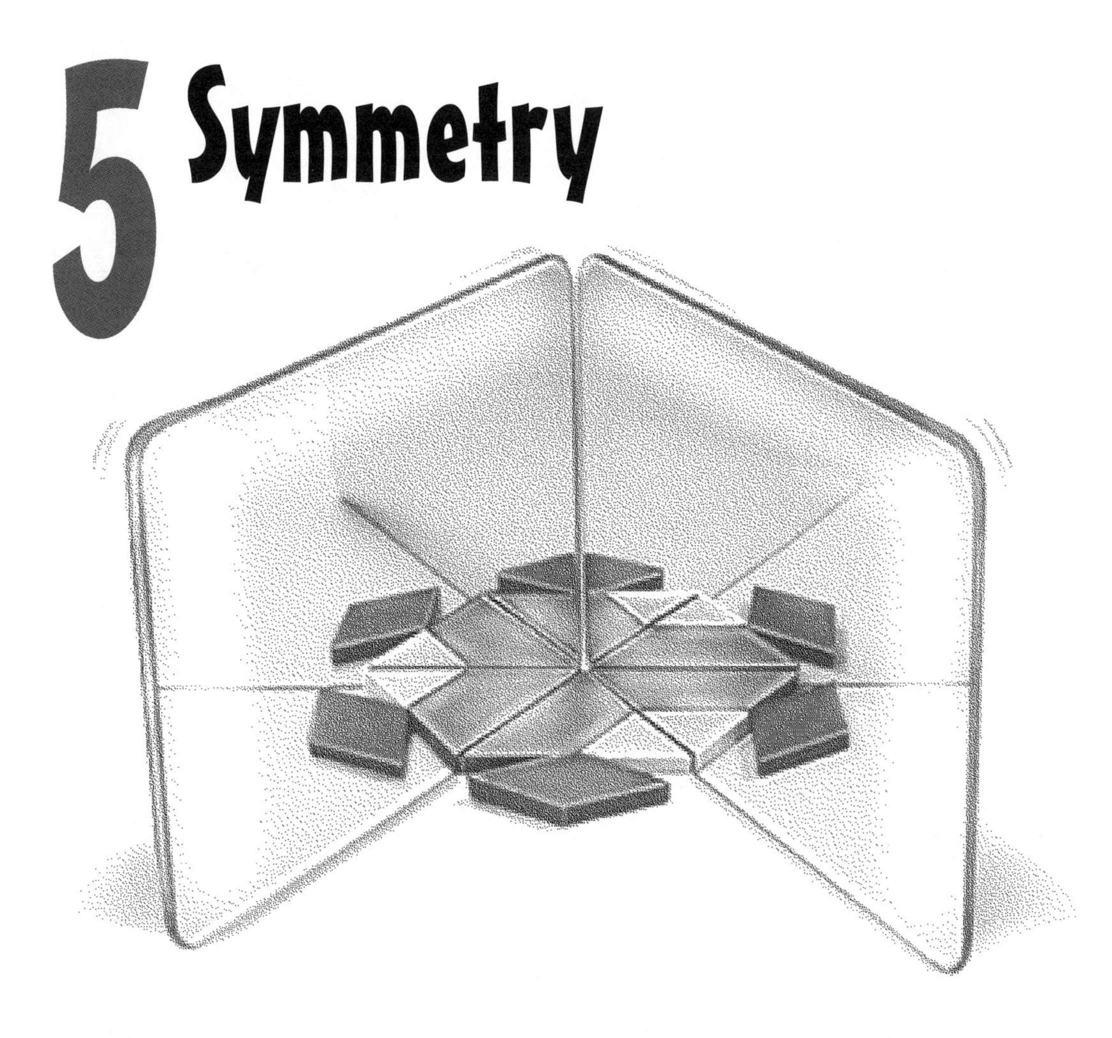

Many of the activities in this section present opportunities for students to express themselves mathematically and artistically—you should be able to create many bulletin board displays from the work they do in various labs. Perhaps you should plan to put together a final exhibit based on the best work from throughout this section. (See the teacher notes to Lab 5.5 for ideas on symmetry exhibits.)

Note that the first lab is a prerequisite for all the other ones.

An excellent resource on the mathematics of symmetry and on crystallographer's notation for symmetric patterns is *Handbook of Regular Patterns: An Introduction to Symmetry in Two Dimensions* by Peter S. Stevens. In addition to a very clear exposition of the ideas, the book includes thousands of symmetric images from just about every culture on the planet. Martin Gardner's *The Ambidextrous Universe* is another good reference on symmetry. It opens with the question, "Why do mirrors reverse left and right, but not up and down?"

See page 191 for teacher notes to this section.

LAB 5.1
Introduction to Symmetry

Name(s) ______________________

1. These letters are *line symmetric*:

A D H I M O T U V W X Y

These letters are not:

F G J L N P R S Z

a. Explain the difference.

b. Draw the line or lines of symmetry on each letter in the first group. (In some cases, there are two lines.)

2. Show how each of the five capital letters not shown in Problem 1 can be written two ways: line symmetric, or not.

3. Organize the lowercase letters in two lists below: line symmetric, and not.

Line symmetric:

Not line symmetric:

4. Below are some figures that have rotation symmetry and some figures that do not. Explain the difference.

5. Organize the capital letters in two lists below: those with rotation symmetry and those without.

Rotation symmetry:

No rotation symmetry:

6. Organize lowercase letters in two lists below: those with rotation symmetry and those without.

Rotation symmetry:

No rotation symmetry:

7. Artist Scott Kim finds ingenious ways to adapt letters so that he can design symmetric words. Describe the symmetry of the design below.

From *Inversions,* by Scott Kim.

8. On a separate sheet of paper, write symmetric words. In this case, symmetry can be interpreted in several ways, using capitals or lowercase letters. The following list gives possible interpretations.

a. Words that have a horizontal line of symmetry, such as COB.

b. Words that have a vertical line of symmetry, such as TOT. (To make this easier, you may write words vertically, with each letter beneath the previous one.)

c. Pairs of words that are mirror images of each other, such as box and pox (using a horizontal mirror).

d. Words that can be read upside down, such as SOS.

e. Pairs of words where one word is the other word upside down, such as MOM and WOW.

f. Palindromes, such as RADAR or RACE CAR (ignoring the space).

Name(s) ______________________

Discussion

A. Line symmetry is also called *mirror* symmetry, *reflection* symmetry, *bilateral* symmetry, and *flip* symmetry. Explain why each of these words is appropriate.

B. What happens if a line-symmetric figure is folded along the line of symmetry?

C. For the figures in Problem 4, draw the lines of symmetry across the figures that are line symmetric, and draw the center of symmetry in those that have rotation symmetry. Note that some of the figures have both kinds of symmetry.

D. The figures in Problem 4 with rotation symmetry include examples of 2-fold, 3-fold, 4-fold, 5-fold, and 6-fold rotation symmetry. Classify each figure according to the specific type of rotation symmetry it has. Explain what it means for a figure to have "*n*-fold" rotation symmetry.

E. If a figure looks unchanged when you rotate it around a point by 180°, it has a special kind of rotation symmetry that is also called *half-turn* symmetry, *central* symmetry, and *point* symmetry. Explain why each of these words is appropriate.

F. What is your definition of the word *symmetry*? Compare it with other students' definitions and the dictionary definition.

G. Which letters have both line and rotation symmetry? What else do they have in common?

H. Some letters are ambiguous and can be written either with or without line symmetry. Others can be written either with or without rotation symmetry.

- Draw and explain examples of these ambiguities.
- Explore how different fonts accentuate or detract from the symmetry of individual letters.

LAB 5.2

Name(s) ______________________

Triangle and Quadrilateral Symmetry

Equipment: Dot or graph paper, isometric dot or graph paper

Eight types of triangles

Equilateral (EQ)
Right isosceles (RI)
Acute scalene (AS)
Half-equilateral (HE)
Acute isosceles (AI)
Obtuse isosceles (OI)
Right scalene (RS)
Obtuse scalene (OS)

Eight types of quadrilaterals

Square (SQ): a regular quadrilateral
Rhombus (RH): all sides equal
Rectangle (RE): all angles 90°
Parallelogram (PA): two pairs of parallel sides
Kite (KI): two distinct pairs of consecutive equal sides, but not all sides equal
Isosceles trapezoid (IT): exactly one pair of parallel sides, one pair of opposite equal sides
General trapezoid (GT): exactly one pair of parallel sides
General quadrilateral (GQ): not one of the above

1. Find two examples of each kind of figure listed above, triangles and quadrilaterals, on dot or graph paper. Label them.
2. On each figure, draw the lines of symmetry if there are any.
3. On each figure, mark the center of symmetry if there is any rotation symmetry.
4. Fill out the table below by entering the types of triangles and quadrilaterals in the appropriate spaces (some spaces may be empty; others may have more than one entry).

	Rotation symmetry			
Line symmetry	**None**	**2-fold**	**3-fold**	**4-fold**
No lines				
One line				
Two lines				
Three lines				
Four lines				

 Name(s) ______________________

Discussion

A. Is there a relationship between the numbers of lines of symmetry and the nature of the rotation symmetry? Explain.

B. For each empty cell in the chart in Problem 4, explain why it is empty. Comment on whether any figures exist with that kind of symmetry (not triangles or quadrilaterals).

C. In some cells, there are two or more figures. Do those figures have anything in common besides symmetry? Discuss.

D. You may have heard that "a square is a rectangle, but a rectangle is not necessarily a square." Does the classification of the quadrilaterals by symmetry throw any light on this statement? Can you use the classification by symmetry to find more statements of this type?

LAB 5.3
One Mirror

Name(s) ______________________

Equipment: Mirror, template

Using your template on a piece of unlined paper, draw a triangle. Then place the mirror on it in such a way as to make a triangle or quadrilateral. For example, starting with an equilateral triangle *ABC* it's possible to make a different equilateral triangle, a kite, or a rhombus.

Check off the figures you make in this manner in the table below, and draw the mirror line on the triangle as a record of how you did it. Also indicate which side you are looking from with an arrow. On the table, mark impossible figures with an X.

Figures made	**By using the mirror on**							
Triangles	**EQ**	**AI**	**RI**	**OI**	**AS**	**RS**	**HE**	**OS**
Equilateral (EQ)	✓							
Acute isosceles (AI)	×							
Right isosceles (RI)								
Obtuse isosceles (OI)								
Acute scalene (AS)								
Right scalene (RS)								
Half-equilateral (HE)								
Obtuse scalene (OS)								
Quadrilaterals	**EQ**	**AI**	**RI**	**OI**	**AS**	**RS**	**HE**	**OS**
General								
Kite	✓							
General trapezoid								
Isosceles trapezoid								
Parallelogram								
Rhombus	✓							
Rectangle								
Square								

Name(s) ____________________

Discussion

A. What polygons besides triangles and quadrilaterals can be made with a triangle and a mirror?

B. Which figures on the table cannot be made with a triangle and a mirror? Why?

C. What do all the figures that can be made have in common?

D. Which line-symmetric figures cannot be made? Why?

E. Which figures can be made from any triangle?

F. What is a good strategy for finding a way to create a given shape with a triangle and mirror?

G. How must the mirror be placed on the original triangle so the resulting figure is a triangle? Why?

H. It is never possible to make an acute isosceles triangle from an obtuse isosceles triangle. However, it is sometimes possible to make an obtuse isosceles triangle from an acute one. Why?

LAB 5.4
Two Mirrors

Name(s) ______________________

Equipment: A pair of mirrors hinged with adhesive tape, pattern blocks, template

In this lab, you will investigate what happens to reflections of pattern blocks as you change the angle between the mirrors.

1. Using your mirrors and one blue pattern block, make the figures below. To record your work, draw the mirrors' position on the figures.

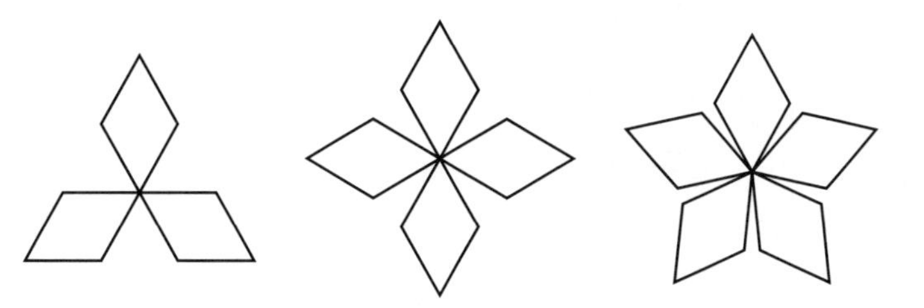

2. For each example in Problem 1, what is the angle between the mirrors? Explain.

3. What is the next figure in the sequence? Make it with a block and mirrors, and draw it in the space at right. Is there another figure after this one?

4. Using a tan pattern block, set up the mirrors so you can see *exactly* 2, 3, 4, 5, and so on copies of it, including the original block (no matter where you look from). For each setup, what is the angle between the mirrors?

5. Calculate the following angles: 360°/3, 360°/4, and so on, up to 360°/12.

6. How can you get the mirrors to form each of the angles you listed in Problem 5 without using a protractor or pattern blocks? (This may not really be possible for the smallest angles. **Hint:** Look at the reflections of the mirrors themselves.)

Name(s) ______________________

7. Which of the angles from Problem 5 yield only mirror-symmetric designs when a pattern block is placed between the mirrors? Which angles can yield either symmetric or asymmetric designs?

   Symmetric only:

   Symmetric or asymmetric:

8. On a separate sheet of paper, create your own designs using one or more pattern blocks and the hinged mirrors. Record your designs with the help of the template. For each design, make a note about the angle you used and whether your design is symmetric.

Discussion

A. Set up the mirrors at an angle of 120°, and place a blue pattern block near one mirror and far from the other, as shown on the figure below. Note that, depending on where you are, you will see two, three, or four blocks (including the original and its reflections). Write 2, 3, and 4 on the figure to indicate where you would look from to see that many blocks.

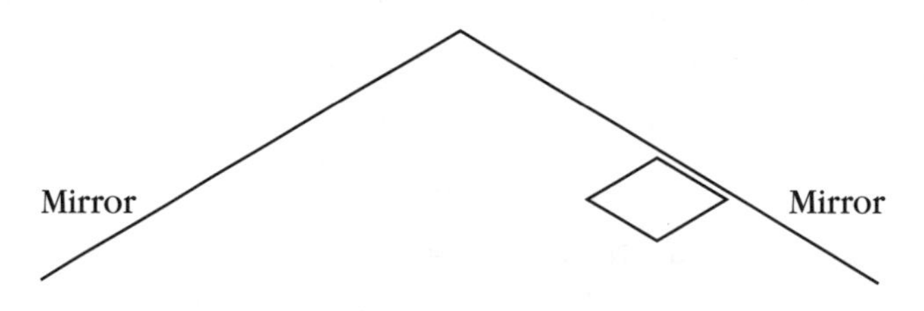

B. Compare the results of using even and odd numbers for n when using the angle $360°/n$ between the mirrors. Discuss both the reflections of the mirrors themselves and the reflections of other objects.

C. Note that, when you use an odd value for n, the middle reflections of the mirrors create the illusion of a two-sided mirror, in the sense that if you look at them from the left or the right you will see a mirror. However, if you look from a point that is equidistant from the two real mirrors, you may see different objects on the left and right of that middle virtual mirror! (Try the setup shown on the figure below.)

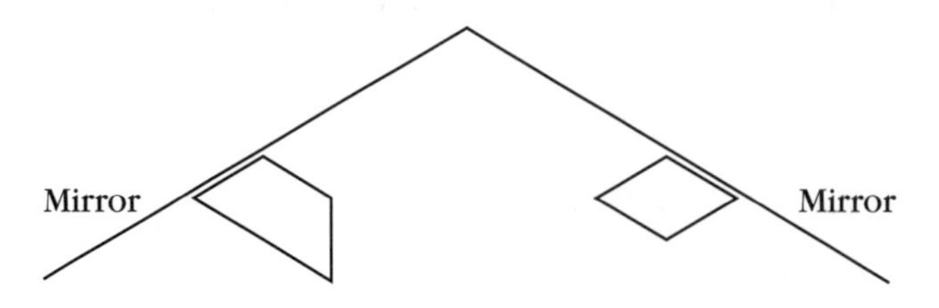

D. If you allow yourself to stand anywhere, what range of angles allows you to see three copies of a block (including the original)? Four copies? Five?

E. Look at the reflection of your face in the mirrors when they make a 90° angle. Wink with your right eye. What happens in the reflection? Can you explain how and why this is different from your reflection in one mirror?

F. Are some of the reflections you see in the hinged mirrors reflections of reflections? Reflections of reflections of reflections? If so, which ones?

G. Write the letter F and look at its reflections in the hinged mirrors. Which copies of it are identical to the original and which are backward?

LAB 5.5
Rotation Symmetry

Name(s) ______________________________

■ **Equipment:** Circle geoboards, Circle Geoboard Paper

1. Make these designs on your circle geoboards.

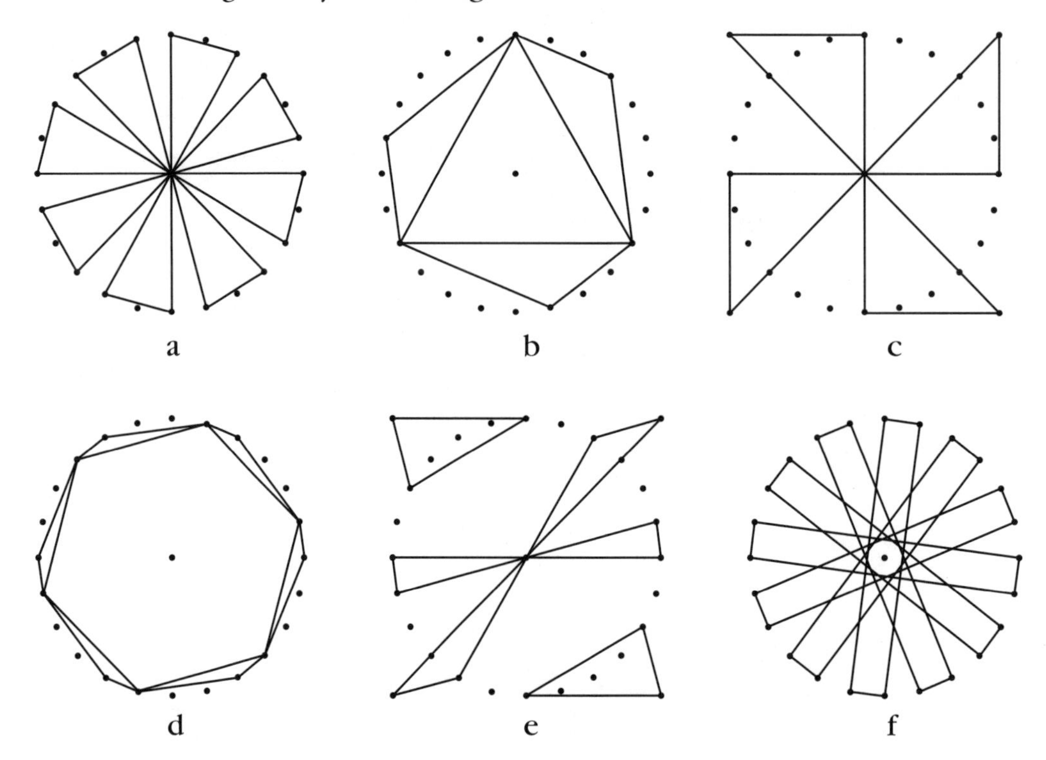

Each of the designs has rotation symmetry: If you rotate the geoboard around its center, the design returns to its initial position before you have done a full turn. For example, figure *b* above would look the same after one-third of a full turn. It returns to its initial position three times as you turn it (counting the final time after the full 360° turn). The design is said to have 3-fold rotation symmetry.

2. Each design in Problem 1 has *n*-fold symmetry for some *n*. What is *n* in each case?

a. ________ **b.** ________ **c.** ________ **d.** ________ **e.** ________ **f.** ________

3. Which of the designs in Problem 1 also have line symmetry? How many lines of symmetry does each one have?

4. On a separate sheet of paper, make designs with *n*-fold rotation symmetry for various values of *n*. Some should have line symmetry and some not. Record and label your designs.

Name(s) ____________________

Discussion

A. It is possible to make the mirror image design for each of the designs in Problem 1. In some cases, the mirror image design will be different from the original one. It is said to have different handedness. Which of the designs have different handedness from their mirror images?

B. Is it possible to have *n*-fold rotation symmetry for any number *n*? What would 1-fold rotation symmetry mean? What about 0-fold? 2.5-fold? –2-fold?

LAB 5.6
Rotation and Line Symmetry

Name(s) ____________________

■ **Equipment:** Pattern blocks, template

Name(s) ______________________

Rotation and Line Symmetry (continued)

Cover the figures on the previous page with pattern blocks. You may use any number and any type of pattern blocks. Each time, draw the resulting figure with the template, and label it with its symmetry properties: If it has rotation symmetry, how many fold? Does it have line symmetry? As you find the various solutions, check them on the table below. If you can make a nonsymmetric design, check the box for 1-fold rotation symmetry. If there is no solution, put an X in the table.

	Triangle		Hexagon		Dodecagon	
	Rot. sym. only	Line sym.	Rot. sym. only	Line sym.	Rot. sym. only	Line sym.
1-fold						
2-fold						
3-fold						
4-fold						
5-fold						
6-fold						
7-fold						
8-fold						
9-fold						
10-fold						
11-fold						
12-fold						

Discussion

A. Which symmetry types are impossible on each of the polygons? Why?

B. What is the relationship between the number of sides of a regular polygon and the possible *n*-fold rotation symmetries once it is covered with pattern blocks? Explain.

C. Describe some strategies for finding new designs from old ones.

D. Given a pattern block figure with a certain symmetry, experiment with making it more symmetric or less symmetric by switching blocks in and out. Keep track of your strategies on a separate sheet of paper.

E. What symmetries do you think would be possible if you tried to cover a figure in the shape of the blue pattern block, with each side equal to 3 inches? Explain your prediction.

LAB 5.7

Name(s) ____________________

Two Intersecting Lines of Symmetry

Equipment: Unlined, graph, or dot paper, template

In this activity, you will create designs based on two intersecting lines of symmetry.

Example: Line 1 and line 2 below form a 45° angle. I started my design by tracing a square in position *a*. In order for the lines to be lines of symmetry of the final figure, I had to add several more squares: *a*, reflected in line 1, gave me *b*; *a* and *b*, reflected in line 2, gave me *c*; *c*, reflected in line 1, gave me *d*; *d*, reflected in line 2, gave me *e*.

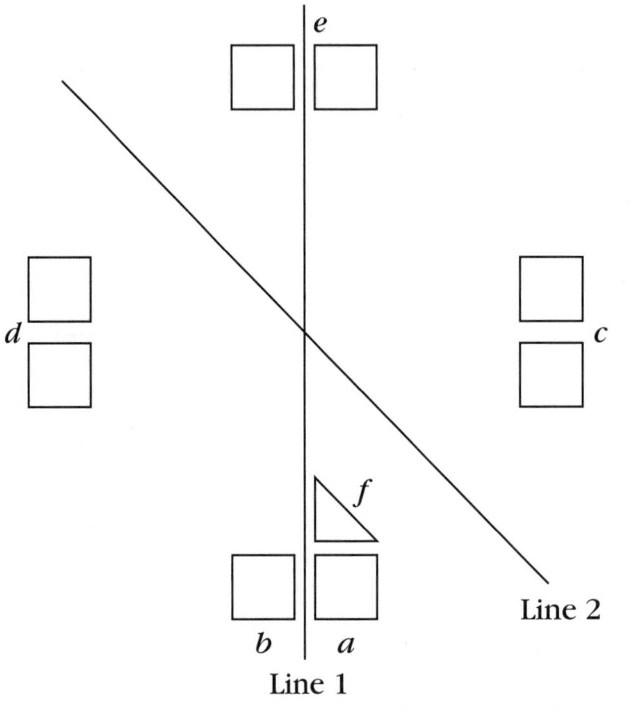

I ended up with eight squares. Any further reflections get me back to previously drawn squares. At that point, I added a triangle (*f*).

1. **a.** Add triangle reflections to the figure, using the lines of symmetry to guide you.

   b. Add anything else you want to the figure, taking care to preserve the symmetry.

2. Create your own designs, using the following angles between the lines of symmetry. (Use the back of this sheet and additional sheets of paper as necessary.)

   a. 90°

   b. 60°

   c. 45°

   d. 30°

3. Create your own designs, using the following angles between the lines of symmetry.

a. 72°

b. 40°

c. 36°

4. Which angles from Problem 3 yield the same designs as each other? Why?

Discussion

A. For each angle, how many copies of any item in the figure are there in the whole figure? (Include the original in the count.)

B. How is this lab similar to the one with the hinged mirror? How is it different?

C. Is it possible to place a physical object (such as a plastic tangram) right on one of the lines of symmetry and have it be part of the design without ruining the symmetry? If not, why not? If yes, how should it be placed? What else would have to be done?

D. Would you answer Question C differently if you were *drawing* an initial shape in such a way that it overlapped a line of symmetry (as opposed to placing a physical object there)?

E. How are a given shape and the reflection of its reflection related? (Reflect first in one, then in the other line of symmetry.) Do they have the same handedness? How could you obtain one shape from the other?

F. While creating these designs, does it make sense to reflect the lines of symmetry themselves? Would it change the final result? If so, how? If not, how does it change the process of getting to the final result?

LAB 5.8

Name(s) ______________________

Parallel Lines of Symmetry

■ **Equipment:** Unlined, graph, or dot paper, templates

In this activity, you will create designs based on two parallel lines of symmetry.

Example: In the figure below, I started with one flag in position *a*. In order for bothe lines to be lines of symmetry of the design, I had to draw several more flags. By reflecting *a* in line 1, I got *b*. By reflecting both of them in line 2, I got *c*. By reflecting *c* in line 1, I got *d*. By reflecting *d* in line 2, I got *e*.

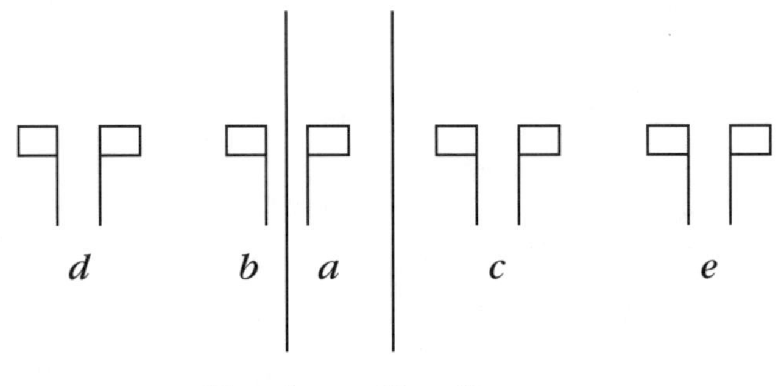

1. Add something to the figure above, taking care to preserve the symmetry.
2. Does this process of adding to the figure ever end? If so, when? If not, why not?
3. On a separate sheet of paper, create your own design based on parallel lines of symmetry.
4. Create your own design based on two parallel lines of symmetry and a third line of symmetry perpendicular to the first two.

The remaining problems are very challenging. Work them out on a separate sheet of paper.

5. Create your own design, based on two parallel lines of symmetry and a third line of symmetry at the following angles.
   a. 60°
   b. 45°
   c. 30°
6. Start with a right isosceles triangle. Extend its sides so they are infinite lines of symmetry, and build a design around them.
7. Repeat Problem 6, starting with the following triangles.
   a. Equilateral triangle
   b. Half-equilateral triangle

Name(s) ____________________

Discussion

A. How could you simulate the pattern of parallel lines of symmetry in the real world?

B. How are a given shape and the reflection of its reflection related? (Reflect first in one, then in the other line of symmetry.) Do they have the same handedness? How could you obtain one shape from the other?

C. While creating these designs, does it make sense to reflect the mirrors themselves? Would it change the final result? If yes, how? If not, how does it change the process of getting to the final result?

D. Describe the fundamental difference between adding a third mirror at 90° and adding it at another angle.

E. Why are the "mirror triangles" in Problems 6 and 7 included in a lesson on parallel lines of symmetry?

6 Triangles and Quadrilaterals

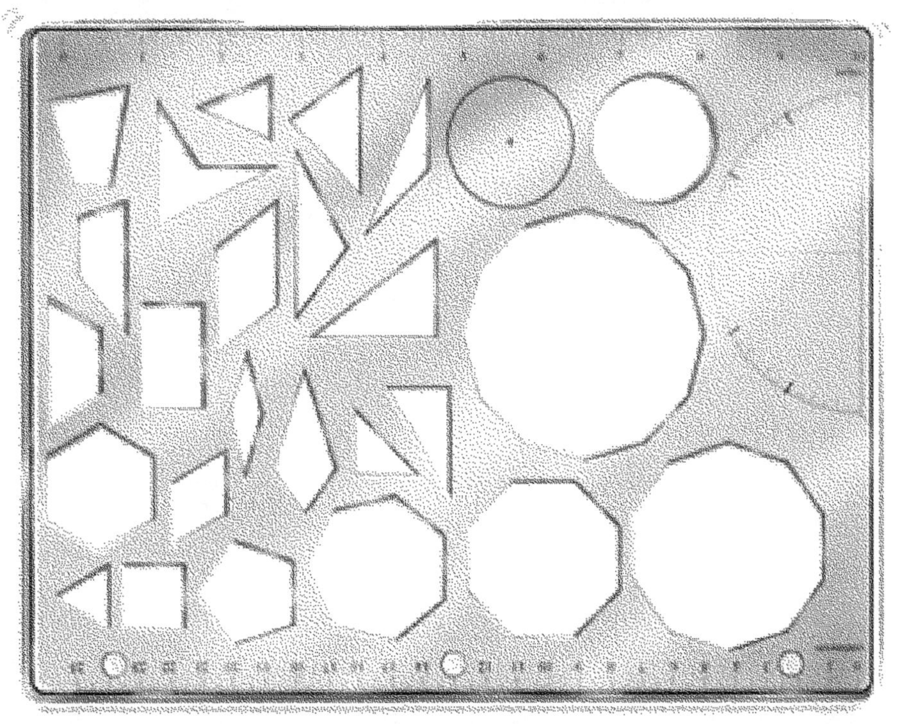

The content of this section is close to some of the topics covered in a traditional high school geometry course. The approach, however, is quite different from that which is followed in most textbooks. The labs work well as discussion starters, and they can be used as supplements to the traditional curriculum before, during, or after the corresponding work on triangles and quadrilaterals.

A note on terminology: I use *equal* when talking about sides or angles and *congruent* when talking about polygons. The distinction between similar and congruent polygons is important to make at this level, and the terminology is helpful there. However, the distinction between congruent sides and equal measurements is rather subtle and really not worth the trouble unless one is developing a very formal axiomatic system. Such formality is often purchased at the price of losing many students and, in fact, the geometric content of the course, as one gets mired in terminology and notation.

See page 199 for teacher notes to this section.

LAB 6.1
Noncongruent Triangles

Name(s) ____________________

Equipment: Compass, straightedge, unlined paper

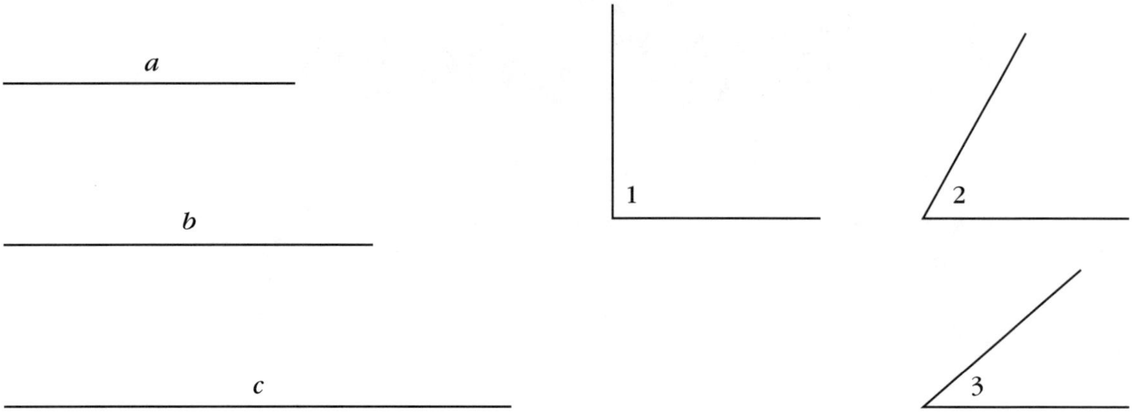

a

b

c

1

2

3

Instructions: For the following problems,

a. Use the compass and straightedge to construct *two noncongruent triangles* that satisfy the given conditions.

b. Label the sides and angles that are equal to those shown above.

c. If you cannot construct two noncongruent triangles, explain why, and construct just one triangle.

d. If you cannot construct even one, explain why.

1. One side of length a and an adjacent angle equal to $\angle 3$
2. One side of length a and an opposite angle equal to $\angle 3$
3. One angle equal to $\angle 1$ and one equal to $\angle 2$
4. One angle equal to $\angle 1$, one equal to $\angle 2$, and one equal to $\angle 3$
5. One side of length a, one of length b, one of length c
6. One side of length a, one of length b, and between them an angle equal to $\angle 3$
7. One side of length a, one of length b, and an angle equal to $\angle 3$ opposite the first side
8. One side of length a between angles equal to $\angle 2$ and $\angle 3$
9. One side of length a adjacent to an angle equal to $\angle 2$ and opposite an angle equal to $\angle 3$
10. One side of length a, one of length b, one of length c, and an angle equal to $\angle 1$

Name(s) ______________________

Discussion

A. Problem 1 could be described as an SA construction (one pair of equal sides and one pair of equal angles). Problem 6 could be described as an SAS construction (two pairs of equal sides with one pair of equal angles between them). Problem 7 could be described as an SSA construction (two pairs of equal sides with one pair of equal angles not between them). Make a two-column table to summarize your work in Problems 1–10.

- In column 1, describe the problem using the letters S and A; each S represents a pair of equal sides, and each A represents a pair of equal angles.
- In column 2, explain whether the given conditions lead to many possible triangles, two possible triangles, one possible triangle, or no triangles.

B. Which problems had no solution? Why?

C. Which problems had a unique solution? How can this help us recognize congruent triangles?

D. Which problem had exactly two solutions?

LAB 6.2 Walking Parallelograms

Name(s) ____________________

The figure at right shows the beginning of a parallelogram walk:

Walk forward one step;

Turn right 140°;

Walk forward three steps.

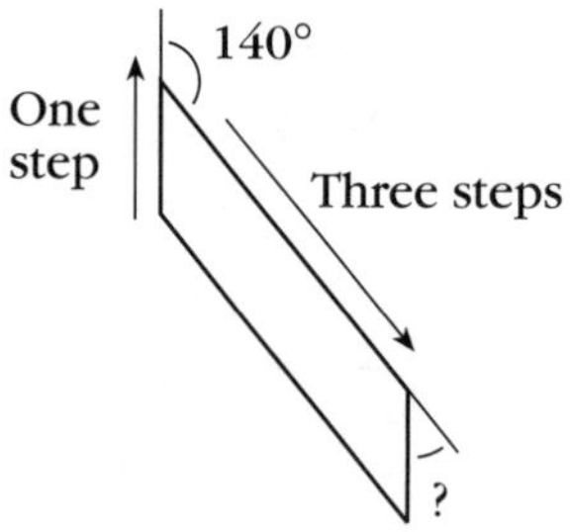

140°
One step
Three steps
?

1. The third side must be parallel to the first. What must the total turning be for the first two turns?

2. Complete the instructions for walking this parallelogram so the walker ends up in the same place and position where he or she started.

3. What are the interior and exterior angles for this parallelogram?

4. Suppose a parallelogram has sides of length x and y and one interior angle measuring $a°$. Write instructions for walking it.

5. Use variables in the same way to write instructions for walking the following figures.

   a. A rhombus

   b. A rectangle

   c. A square

6. Given any rhombus, can you choose values for the variables in the parallelogram instructions so that someone following them would walk that rhombus? Explain.

7. Given any parallelogram, can you choose values for the variables in the rhombus instructions so that someone following them would walk that parallelogram? Explain.

8. Use the answers to Problems 6 and 7 to answer these questions.

a. Is a rhombus a parallelogram?

b. Is a parallelogram a rhombus?

9. Repeat Problems 6 and 7 for other pairs of quadrilaterals among these four: parallelogram, rhombus, rectangle, square. There are four possibilities besides the rhombus–parallelogram pair.

10. Use the answer to Problem 9 to decide which of the four quadrilaterals are specific examples of one of the others.

Discussion

A. What is the relationship between consecutive angles of a parallelogram? Between interior and turn angles?

B. How many variables does each of your sets of instructions in Problems 4 and 5 have? What does this tell you about the corresponding shape?

C. What values are allowed for each variable? What values are not?

D. For which two of the shapes (parallelogram, rhombus, rectangle, square) is neither one an example of the other?

E. How would you define congruence rules for each of these four quadrilaterals? For example, "Two squares are congruent if they have a pair of equal sides (S)."

F. Some math teachers believe that a trapezoid should be defined as a quadrilateral with *exactly* one pair of parallel sides. Others believe that a trapezoid should be defined as a quadrilateral with *at least* one pair of parallel sides. How are these choices different? How do they affect the classification of quadrilaterals? Whom do you agree with and why?

LAB 6.3

Name(s) ______________________

Making Quadrilaterals from the Inside Out

- **Equipment:** Dot or graph paper, straightedge

Definition: To bisect means to cut *exactly* in half.

Instructions: For the following problems,

a. On dot or graph paper, draw two intersecting line segments that satisfy the given conditions. These will be the diagonals.

b. Join the endpoints to make a quadrilateral.

c. Name the quadrilateral.

Whenever possible, try to make a quadrilateral that is not a parallelogram, a trapezoid, or a kite.

1. Perpendicular, equal segments that bisect each other
2. Perpendicular, equal segments that do not bisect each other
3. Perpendicular, unequal segments that bisect each other
4. Perpendicular, unequal segments that do not bisect each other
5. Nonperpendicular, equal segments that bisect each other
6. Nonperpendicular, equal segments that do not bisect each other
7. Nonperpendicular, unequal segments that bisect each other
8. Nonperpendicular, unequal segments that do not bisect each other

Discussion

A. For four of the Problems 1–8, it is possible to get a quadrilateral that is not a parallelogram, a trapezoid, or a kite. What do they all have in common?

B. What requirements do you need to put on the diagonals to get a kite? (It is not one of conditions 1–8.)

C. What requirements do you need to put on the diagonals to get an isosceles trapezoid? (It is not one of conditions 1–8.)

LAB 6.4

Name(s) ______________________

Making Quadrilaterals from Triangles

■ **Equipment:** Template

Eight types of triangles

Equilateral (EQ)
Right isosceles (RI)
Acute scalene (AS)
Half-equilateral (HE)
Acute isosceles (AI)
Obtuse isosceles (OI)
Right scalene (RS)
Obtuse scalene (OS)

Use a separate sheet or sheets of paper to prepare a well-organized and clearly illustrated report that answers the following questions.

1. Using two congruent triangles placed edge to edge, what types of quadrilaterals can you make? For example, as shown below, with two congruent right isosceles triangles, you can make a square or a parallelogram. Discuss all eight types of triangles.

2. Using three congruent triangles placed edge to edge, what types of quadrilaterals can you make?

LAB 6.5
Slicing a Cube

Name(s) ____________________

Equipment: Transparency, stiff paper, scissors, adhesive tape

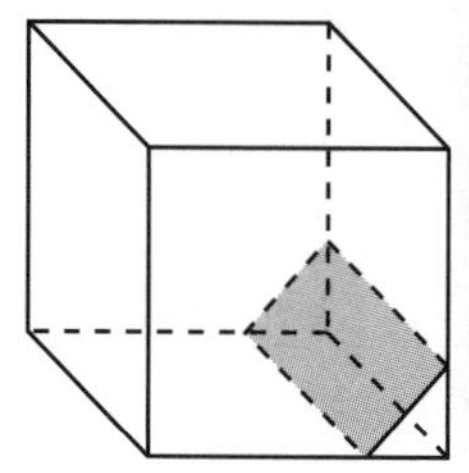

Imagine that you slice a cube in a direction parallel to one of the faces. The shape of the slice will be a square. Now imagine that you slice a cube parallel to one of the bottom edges, but tilted from the horizontal. The shape of the slice will be a rectangle.

What shapes are possible for a slice? To investigate this, you will use a hollow transparent cube and stiff paper to simulate the slices.

1. Cut five squares out of the transparency, each one with sides 3 inches in length. Tape them together to make a lidless cubic box.
2. Cut a 1- by 3-inch rectangle out of stiff paper, and place it in the cube so as to match the slice shown in the figure above.
3. Cut other shapes out of stiff paper to show slices that match each of the following polygons, if possible. Sketch your solutions on a separate sheet of paper.

Triangles:	Quadrilaterals:	Other polygons:
Equilateral (EQ)	Square (SQ)	Pentagon
Acute isosceles (AI)	Rhombus (RH)	Regular pentagon
Right isosceles (RI)	Rectangle (RE)	Hexagon
Obtuse isosceles (OI)	Parallelogram (PA)	Regular hexagon
Acute scalene (AS)	Kite (KI)	Seven-gon
Right scalene (RS)	Isosceles trapezoid (IT)	
Half-equilateral (HE)	General trapezoid (GT)	
Obtuse scalene (OS)	General quadrilateral (GQ)	

Discussion

A. What polygon slices are possible? Impossible? Explain.

B. What is the range of possible sides and angles for each polygon slice?

C. What is the most symmetric triangle slice? Quadrilateral? Pentagon? Hexagon?

D. Is there a square slice that is not parallel to one of the faces?

7 Tiling

In this section, we address one question from several points of view: What are some ways to tile a plane? *To tile a plane* means to cover an infinite flat surface with an unlimited supply of a finite number of shapes, with no gaps or overlaps. Some types of tilings are also called *tessellations*. To keep jargon to a minimum, we will use *tiling* rather than *tessellation*.

Some of the resulting investigations are directly connected with the traditional high school geometry curriculum and reinforce standard topics, particularly Lab 7.3 (Tiling with Triangles and Quadrilaterals) and Lab 7.4 (Tiling with Regular Polygons).

Every lab in this section lends itself to the creation of a bulletin board display of student work, featuring the best-looking tilings and some explanations of the underlying mathematics.

You may not have enough time to do all four labs on this topic, or you may suspect that your students would find them to be too much on the same topic. If so, you may use Lab 7.1 (Tiling with Polyominoes) to introduce the concepts and distribute the remaining labs among your students, or groups of students, as independent projects, which can be combined in a final class exhibit. Many students are fascinated with M. C. Escher's graphic work, which involves tiling with figures such as lizards, birds, or angels. A report on Escher could be one of the independent projects.

Note that tilings provide interesting infinite figures for students to analyze from the point of view of symmetry. To support and encourage that inquiry, the same symmetry questions are asked about each tiling in each discussion section. That kind of analysis can be a worthwhile complement to the projects suggested above.

Although tiling is as ancient as tiles, it is a fairly recent topic in mathematics. Except for Kepler, all of the research on it has occurred since the nineteenth century, and new results continue to appear today. The definitive reference on the mathematics of tiling is *Tilings and Patterns* by Banko Grunbaum and G. C. Shephard (Freeman, N.Y., 1987), which features many beautiful figures.

See page 203 for teacher notes to this section.

LAB 7.1
Tiling with Polyominoes

Name(s) ______________________

■ **Equipment:** Grid paper

Imagine a flat surface, extending forever in all directions. Such a surface is called a *plane*. If you had an unlimited supply of rectangular polyomino tiles, you could cover as large an area as you wanted. For example, you could lay out your tiles like this:

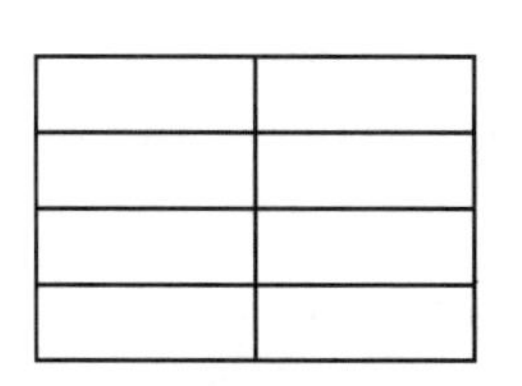

. . . or like this:

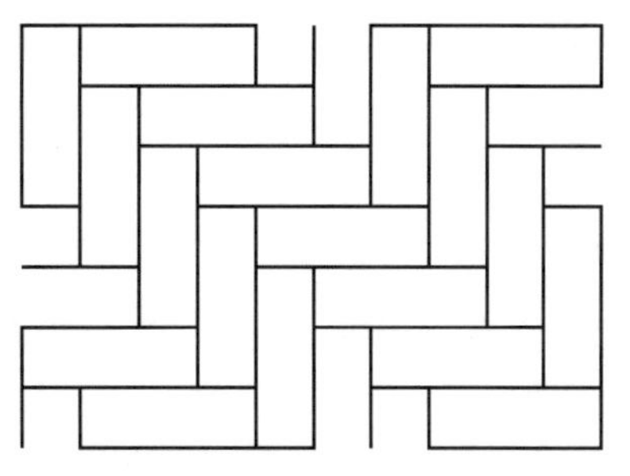

1. Find interesting ways to tile a plane with rectangular polyominoes. Record them on your grid paper.

Below are two bent tromino tilings:

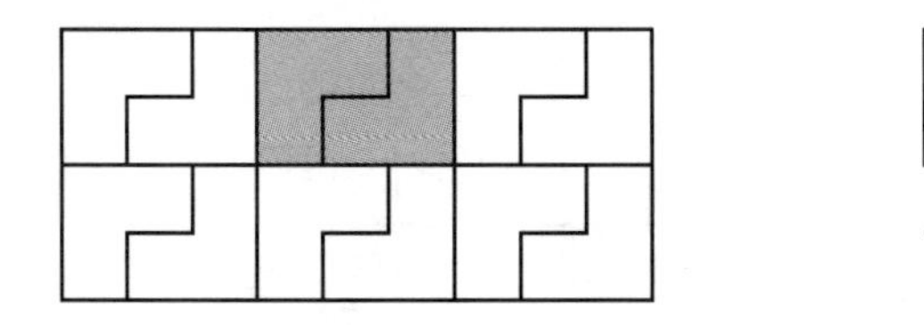

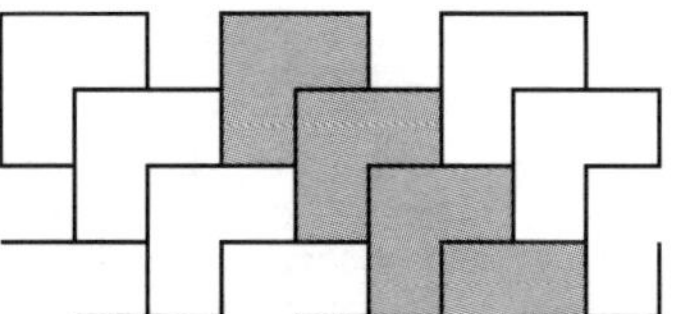

The first one is made of two-tromino rectangles; the second is made of diagonal strips. By extending the patterns, you can tile a plane in all directions.

2. Can you extend this U pentomino tiling in all directions? Experiment and explain what you discover.

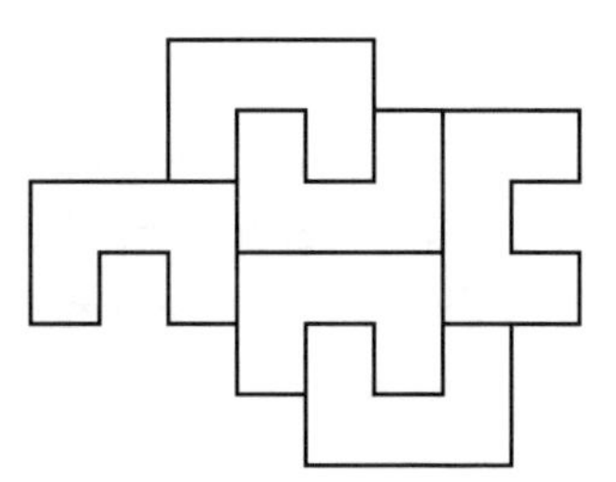

3. Find ways to tile a plane with each of the following tetrominoes.
   a. n
   b. l
   c. t
4. Find ways to tile a plane with each of the twelve pentominoes.

Name(s) ______________

Discussion

A. Would it be possible to find tilings for rectangles that are not polyominoes? (In other words, their dimensions are not whole numbers.)

B. How can you be sure a given tiling will work on an infinite plane? Try to establish general rules for answering this question, and summarize them in a paragraph.

C. In Problems 3 and 4, you can add the constraint that the polyominoes cannot be turned over, just rotated and moved over. Is it possible to tile a plane under these conditions?

D. Polyomino tilings may involve the polyomino in one or more positions. For example, the first figure in this lab shows a polyomino that is always in the same position. The second shows a polyomino in two different positions. Can you set additional challenges for each polyomino by looking for tilings where it appears in one, two, three, or more different positions?

E. The figures below show two- and three-polyomino tilings. Create your own.

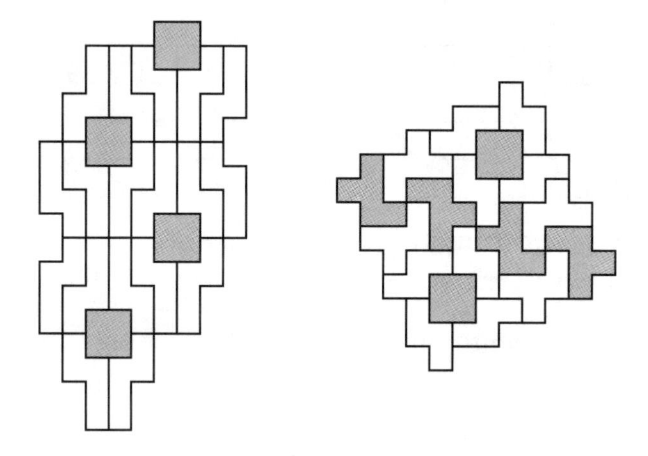

F. Analyze the symmetries of the tilings that you found. Do they have lines of symmetry? Are all of those parallel to each other? If not, what angle do they make with each other? Do the tilings have centers of rotation symmetry? How are those arranged? Do all the centers have the same n-fold symmetry?

LAB 7.2
Tiling with Pattern Blocks

Name(s) ________________________

Equipment: Pattern blocks, unlined paper, template

It is easy to tile a plane with pattern blocks. Below are two possible orange tilings:

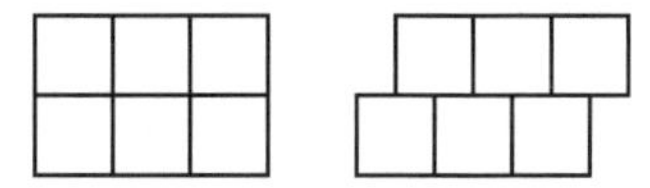

In the first one, all the vertices are o90-o90-o90-o90. In the second, all the vertices are o90-o90-o180. (The o stands for *orange;* the numbers represent the measures in degrees of the angles surrounding the vertex.)

1. Find at least two one-color pattern block tilings. Use your template to draw them on unlined paper. For each one, describe the vertices as in the examples above.
2. In some pattern block tilings, there are different types of vertices. For example, in the one at right, there are two: g60-g60-g60-g60-g60-g60 and g60-g60-o90-g60-o90. Mark the two types of vertices using different-colored dots on the figure.

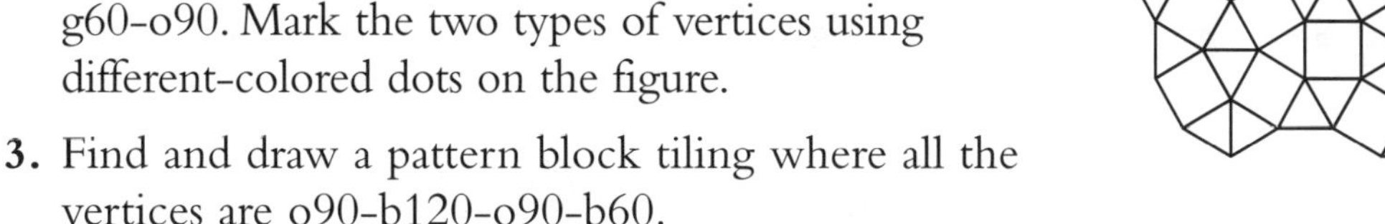

3. Find and draw a pattern block tiling where all the vertices are o90-b120-o90-b60.
4. Find and draw at least two other two-color tilings. For each one, describe the vertices as above.
5. Find and draw one pattern block tiling with the following types of vertices.
   a. o90-t150-b120
   b. o90-b60-t30-o90-b60-t30
6. Find another three-color pattern block tiling. Describe its vertices.
7. Explain why there could not be a vertex of type r60-b120-r120 in a tiling of a plane.

Discussion

A. What must be true of the numbers in the description of a vertex? How can that help us discover new pattern block tilings?

B. Find a way to make the vertex notation more concise. Do you think that a more concise notation would be an improvement?

C. Analyze the symmetries of the tilings that you found. Do they have lines of symmetry? Are all of those parallel to each other? If not, what angle do they make with each other? Do the tilings have centers of rotation symmetry? How are those arranged? Do all the centers have the same *n*-fold symmetry?

LAB 7.3

Name(s) ____________________

Tiling with Triangles and Quadrilaterals

- **Equipment:** Template, unlined paper

Eight types of triangles

Equilateral (EQ)
Right isosceles (RI)
Acute scalene (AS)
Half-equilateral (HE)
Acute isosceles (AI)
Obtuse isosceles (OI)
Right scalene (RS)
Obtuse scalene (OS)

Eight types of quadrilaterals

Square (SQ)
Rectangle (RE)
Kite (KI)
General trapezoid (GT)
Rhombus (RH)
Parallelogram (PA)
Isosceles trapezoid (IT)
General quadrilateral (GQ)

1. Using your template on unlined paper, work with your classmates to find a way to tile a plane with each of the figures listed above.
2. Can you find a triangle that cannot tile a plane? If yes, explain how you know it will not tile. If you think any triangle can tile, explain how to do it, basing your explanation on a scalene triangle.
3. Tile a plane with a nonconvex quadrilateral.
4. Can you find a quadrilateral that cannot tile a plane? If so, explain how you know it will not. If you think any quadrilateral can tile a plane, explain how to do it, basing your explanation on a general quadrilateral.

Discussion

A. In some cases, there may be different tilings based on the same tile. Find variations on some of the sixteen tilings you found in Problem 1.

B. A parallelogram tiling can be turned into a scalene triangle tiling by drawing the diagonals of the parallelograms. In the other direction, some scalene triangle tilings can be turned into parallelogram tilings by combining pairs of triangles into parallelograms. Find other relationships of this type among tilings.

C. Analyze the symmetries of the tilings that you found. Do they have lines of symmetry? Are all of those parallel to each other? If not, what angle do they make with each other? Do the tilings have centers of rotation symmetry? How are those arranged? Do the centers all have the same n-fold symmetry?

LAB 7.4 Name(s) ______________________
Tiling with Regular Polygons

■ **Equipment:** Template, pattern blocks, unlined paper

1. Which regular polygons can be used to tile a plane? Draw the tilings on unlined paper using your template. (Multiple solutions are possible in some cases.)
2. Explain why other regular polygons would not work.

3. Find and draw a tiling that uses two different regular polygons.

In order to tile a plane with polygons, each vertex must be surrounded by angles that add up to 360°. This observation leads to a method for finding ways to tile by seeing what regular polygons can be used to surround a point. For example, two octagons and one square do work.

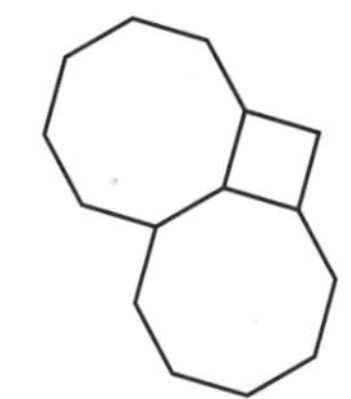

4. Find as many ways as possible to place regular polygons from your template around a point with one vertex of each polygon touching the point. List the ways you find below.

5. Some of the answers to Problem 4 lead to possible tilings of a plane. Find and draw some of them.

6. Some of the answers to Problem 4 do not lead to a possible tiling of a plane. Choose one and explain why it does not work.

Name(s) ____________________

Discussion

A. Some regular polygons, or combinations of regular polygons, can be put together to make smooth strips (infinite strips with straight edges). Those can then be combined in any number of ways to make tilings of a plane. Find such strips.

B. There are five other ways to use regular polygons to surround a point. They involve a 15-gon, an 18-gon, a 20-gon, a 24-gon, and a 42-gon. In each of the five cases, find the other regular polygons that are needed.

C. With the additional conditions that tiles can only be placed edge to edge and all vertices are identical, there are only eleven possible regular polygon tilings. Find as many as you can.

D. Analyze the symmetries of the tilings that you found. Do they have lines of symmetry? Are all of those parallel to each other? If not, what angle do they make with each other? Do the tilings have centers of rotation symmetry? How are those arranged? Do the centers all have the same *n*-fold symmetry?

8 Perimeter and Area

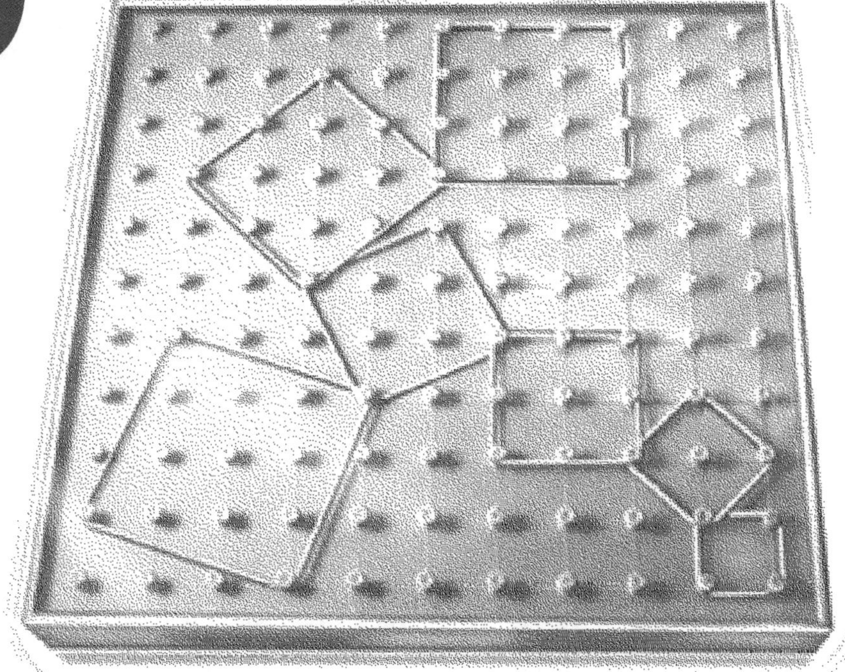

This section starts a discussion of perimeter and area, which will be continued throughout the rest of this book. Contexts involving square roots and the Pythagorean theorem will be taken up in Section 9. Contexts involving similarity and scaling will come up in Section 10. And finally, contexts involving trigonometry will be taken up in Section 11.

Some of this material is explored (with an emphasis on connections to algebra) in the textbook I coauthored: *Algebra: Themes, Tools, Concepts* (Creative Publications, 1994).

See page 208 for teacher notes to this section.

LAB 8.1
Polyomino Perimeter and Area

Name(s) ______________________

■ **Equipment:** 1-Centimeter Grid Paper, interlocking cubes

A polyomino is a graph paper figure whose outline follows the graph paper lines and never crosses itself. We will only consider polyominoes with no holes, such as these:

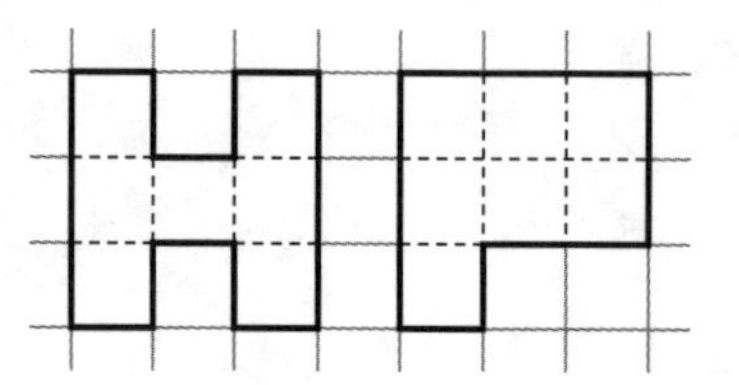

1. What are the area and perimeter of each polyomino shown above?

2. Find a polyomino with the same area as the ones above, but with a different perimeter. Sketch it in the grid below.

 Name(s) ____________

Polyomino Perimeter and Area (continued)

3. Experiment on graph paper or with the help of your interlocking cubes and fill out the following table.

Area	Minimum perimeter	Maximum perimeter
1		
2		
3	8	8
4		
5	10	12
6		
7		
8		
9		
10		
11		
12		
13		

Area	Minimum perimeter	Maximum perimeter
14		
15		
16		
17		
18		
19		
20		
21		
22		
23		
24		
25		
26		

4. Find a formula for the maximum perimeter, P_{max}, for a given area A.

5. Describe a pattern for the minimum perimeter.

6. What would the minimum and maximum perimeters be for the following areas?

a. 49 min. ________ max. ________

b. 45 min. ________ max. ________

c. 50 min. ________ max. ________

d. 56 min. ________ max. ________

Name(s) ____________________

Discussion

A. Is it possible to draw a polyomino with an odd perimeter? Explain how to do it, or why it is impossible.

B. While filling out the table, what was your strategy for finding the maximum and minimum perimeters?

C. Graph minimum and maximum polyomino perimeters with area on the x-axis and perimeter on the y-axis. What does the graph show about the formula and pattern you found in Problems 4 and 5?

D. Find the polyominoes whose area and perimeter are numerically equal.

E. Explain your strategy for answering the questions in Problem 6.

LAB 8.2

Name(s) ____________________

Minimizing Perimeter

Equipment: 1-Centimeter Grid Paper, the results of Lab 8.1

In this lab, you will try to get a better understanding of the minimum perimeter pattern you discovered in Lab 8.1.

By studying the numbers in the table in Lab 8.1, you will notice that when the area increases by 1, the minimum perimeter stays the same or increases by 2.

1. Thinking about them numerically, what is special about the areas that precede increases of the minimum perimeter? **Hint:** There are two different types of such areas.

2. Circle all perfect square areas in the table in Lab 8.1.

3. Circle all areas that are the products of consecutive whole numbers. For example, circle 6, since $6 = 2 \cdot 3$. (We will call these products *rectangular numbers*.)

Square and rectangular numbers are the key to predicting the minimum perimeter for any area. For example, consider the square number 900.

4. What are the minimum perimeters for the following areas?

   a. 900

   b. 901

   c. 895

5. What is the greatest rectangular number that is less than 900? What is the least rectangular number that is greater than 900?

Name(s) ____________________

6. What are the minimum perimeters for the following areas?

a. 925

b. 935

c. 865

In order to understand the geometry behind the special role of square and rectangular numbers, you can draw consecutive minimum-area polyominoes by spiraling around an initial square, as shown in the figure below.

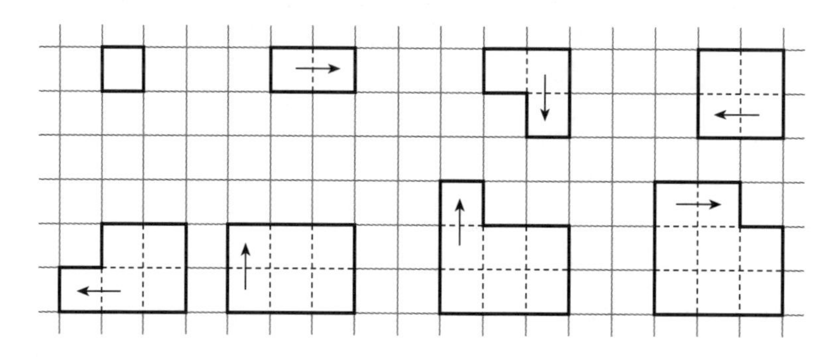

7. Continue the process shown in the figure. Describe situations when the perimeter stays the same, and when it changes.

Discussion

A. Explain why square and rectangular numbers are important in understanding this lab.

B. Explain, with the help of figures, why the perimeter of a polyomino whose area is a little less than a given square or rectangular number is the same as the perimeter of the square or rectangle.

C. What are the minimum perimeters of polyominoes with the following areas?

a. n^2

b. $n(n + 1)$

LAB 8.3

Name(s) ______________________

A Formula for Polyomino Perimeter

■ **Equipment:** 1-Centimeter Grid Paper, interlocking cubes

It is possible to find a formula for the perimeter of a polyomino as a function of its area and of one other variable: the number of *inside dots.*

Inside dots are the points of intersection of grid lines in the interior of the polyomino. For example, the polyomino at right has three inside dots.

1. What are the perimeter, the area, and the number of inside dots for the figures below?

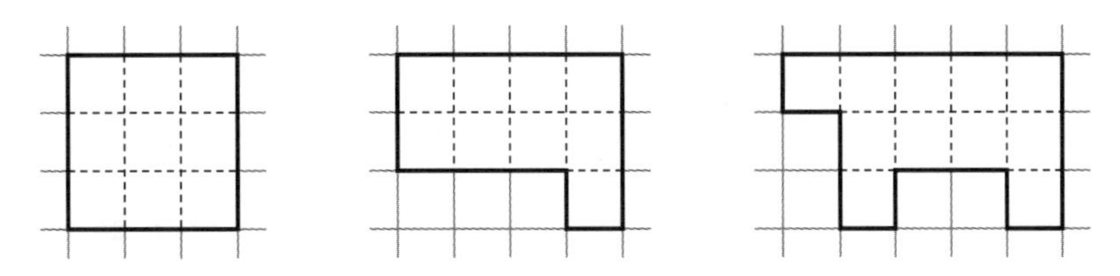

Experiment on grid paper or with your interlocking cubes to find the formula. Keep track of your experiments in the table. Problems 2–5 can help you organize your research.

Perimeter	Area	Inside dots

Perimeter	Area	Inside dots

2. Create a series of figures, all with the same area but with different numbers of inside dots. Keep a record of the area, perimeter, and number of inside dots in the table.
3. Every time you add an inside dot, how are you changing the perimeter?
4. Create a series of figures, all with the same number of inside dots but with different areas. Keep a record of the area; perimeter, and inside dots in the table.
5. Every time you add one square unit to the area, how are you changing the perimeter?
6. Find a formula for the perimeter P of a polyomino if its area is A and it has i inside dots.
7. **Puzzle:** Using interlocking cubes, make all possible polyominoes with perimeter 10. Use only one color per polyomino. Arrange them in a rectangle and record your solution below.

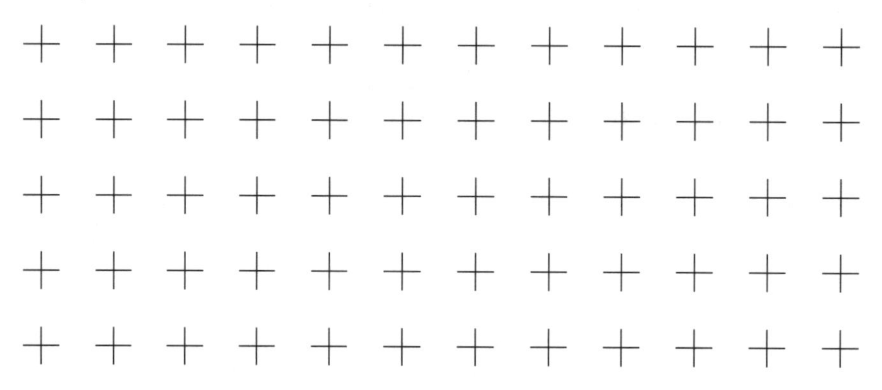

Discussion

A. If two polyominoes have the same area and the same perimeter, what can you say about the number of inside dots?

B. If two polyominoes have the same area, but the perimeter of one is greater than the perimeter of the other, what can you say about their respective numbers of inside dots?

C. What is the least number of inside dots for a given area?

D. What is the greatest area for a given number of inside dots?

E. What can you say about the number of inside dots for a square of side n?

F. What can you say about the number of inside dots for a rectangle of dimensions n and $n + 1$?

LAB 8.4
Geoboard Area

Name(s) ____________________

■ **Equipment:** Geoboard, dot paper

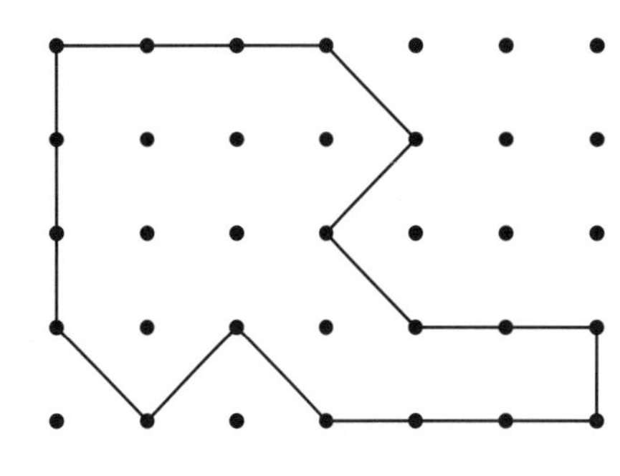

The area of the geoboard figure at right is 15.

1. Find other geoboard figures with area 15. The boundaries of the figures need not be horizontal or vertical. Find figures that are different from the ones your neighbors find. Record your solutions on dot paper.

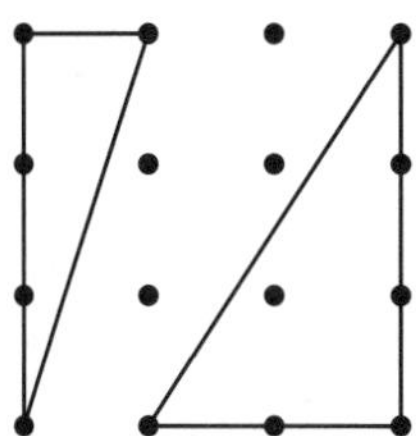

It is easiest to find areas of geoboard rectangles with horizontal and vertical sides. The next easiest figures are the "easy" right triangles, such as the two shown at right.

2. Find the areas of these triangles.

If you can find the area of easy right triangles, you can find the area of any geoboard figure!

3. Find the areas of the figures below.

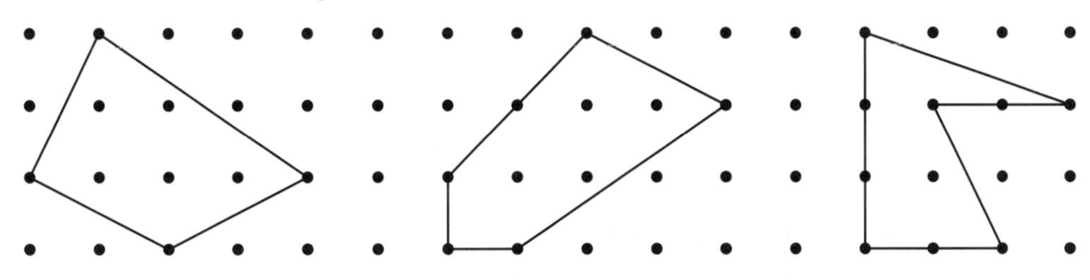

4. Find the area of the figure below. (It may be more difficult than Problem 3. **Hint:** Use subtraction.)

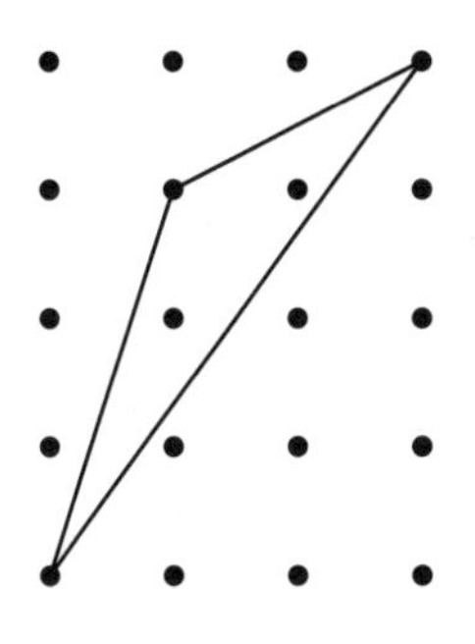

LAB 8.4

Name(s) ______________________

Geoboard Area (continued)

5. Find the areas of the figures below.

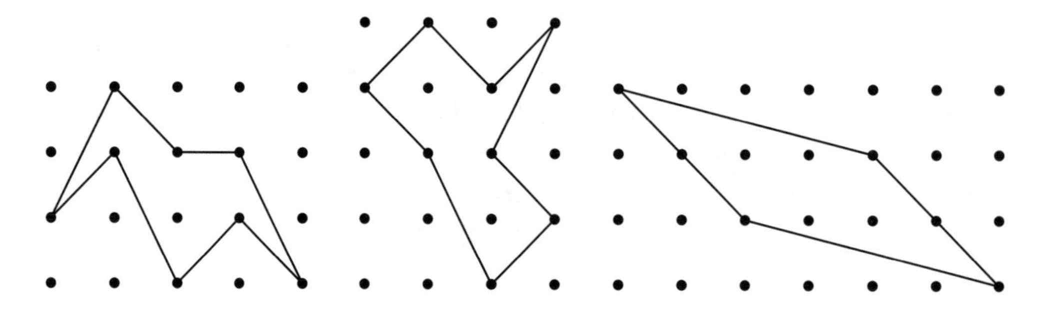

6. How many noncongruent geoboard triangles are there with area 8? Limit yourself to triangles that can be shown on an 11 × 11 geoboard and have a horizontal base. Record your findings on dot paper.

7. **Puzzle:** Find the geoboard figure with the smallest area in each of these categories.

   a. Acute triangle
   b. Obtuse triangle
   c. Right triangle
   d. Square
   e. Rhombus (not square)
   f. Rectangle (not square)
   g. Kite
   h. Trapezoid
   i. Parallelogram

Discussion

A. A common mistake in finding geoboard areas is to overestimate the sides of rectangles by 1 (for example, thinking that the rectangle at right is 4 × 5). What might cause this mistake?

B. Explain, with illustrated examples, how the following operations may be used in finding the area of a geoboard figure: division by 2; addition; subtraction.

C. What happens to the area of a triangle if you keep its base constant and move the third vertex in a direction parallel to the base? Explain, using geoboard or dot paper figures.

D. Use geoboard figures to demonstrate the area formulas for various quadrilaterals.

LAB 8.5
Geoboard Squares

Name(s) ______________________

■ **Equipment:** Geoboard, dot paper

1. There are 33 different-size squares on an 11 × 11 geoboard. With the help of your neighbors, do the following.

a. Find all the squares.

b. Sketch each square on dot paper, indicating its area and the length of its side.

Discussion

A. How can you make sure that two sides of a geoboard square really form a right angle?

B. How can you organize your search so as to make sure you find all the squares?

C. Is it possible to find squares that have the same area, but different orientations?

D. In the figure at right, find the following in terms of a and b.

a. The side of the outside square

b. The area of the outside square

c. The area of each triangle

d. The area of the inside square

e. The side of the inside square

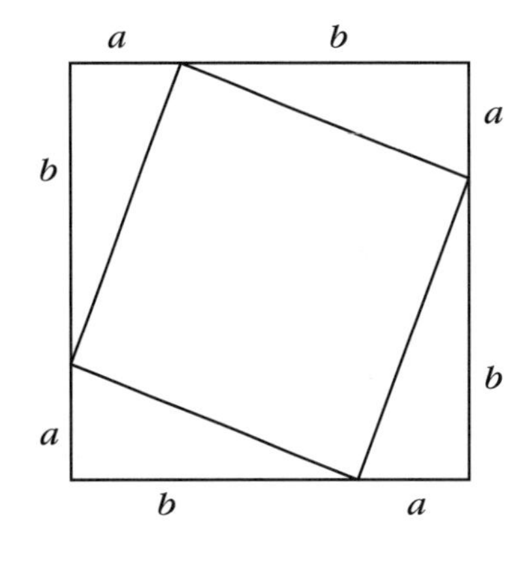

LAB 8.6
Pick's Formula

Name(s) ____________________

Equipment: Geoboard, dot paper

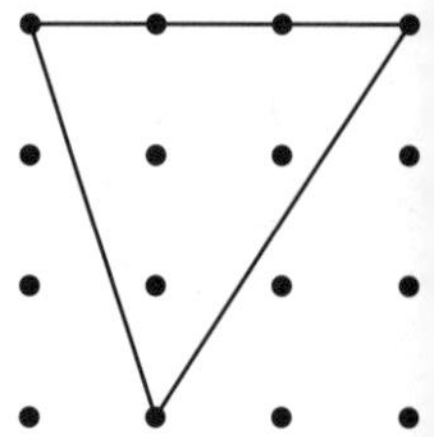

It is possible to find a formula for geoboard area as a function of boundary dots and inside dots.

For example, the geoboard figure at right has three inside dots and five boundary dots.

1. What is the area of the figure above?
2. What are the area, the number of inside dots, and the number of boundary dots for each of the figures at right?

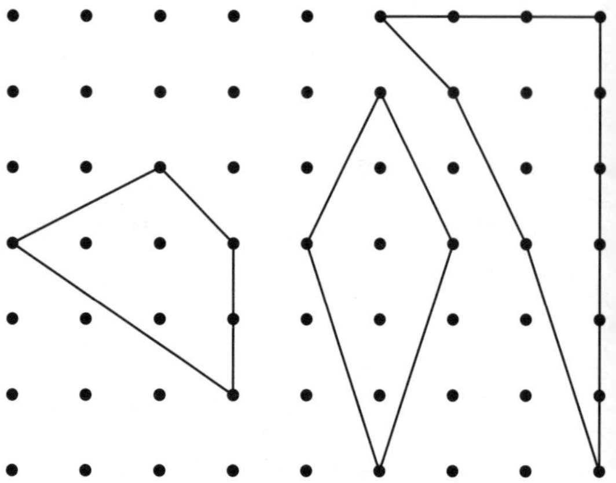

Experiment on your geoboard or on dot paper to find the formula. Keep track of your experiments in the table below. Problems 3–6 can help you organize your research.

Inside dots	Bound. dots	Area

Inside dots	Bound. dots	Area

Inside dots	Bound. dots	Area

3. Create a series of geoboard figures, all with the same number of boundary dots but with different numbers of inside dots. Keep a record of the area, boundary dots, and inside dots in the table.

4. Every time you add an inside dot, how are you changing the area?

5. Create a series of figures, all with the same number of inside dots but with different numbers of boundary dots. Keep a record of the inside dots, boundary dots, and area in the table.

6. Every time you add 1 to the number of boundary dots, how are you changing the area?

7. Find a formula for the area of a geoboard figure if it has b boundary dots and i inside dots.

Discussion

A. What is the greatest area for a given number of inside dots?

B. What is the least area for a given number of boundary dots?

C. How is the formula you found in this lab related to the one you found in Lab 8.3 (A Formula for Polyomino Perimeter)?

9 Distance and Square Root

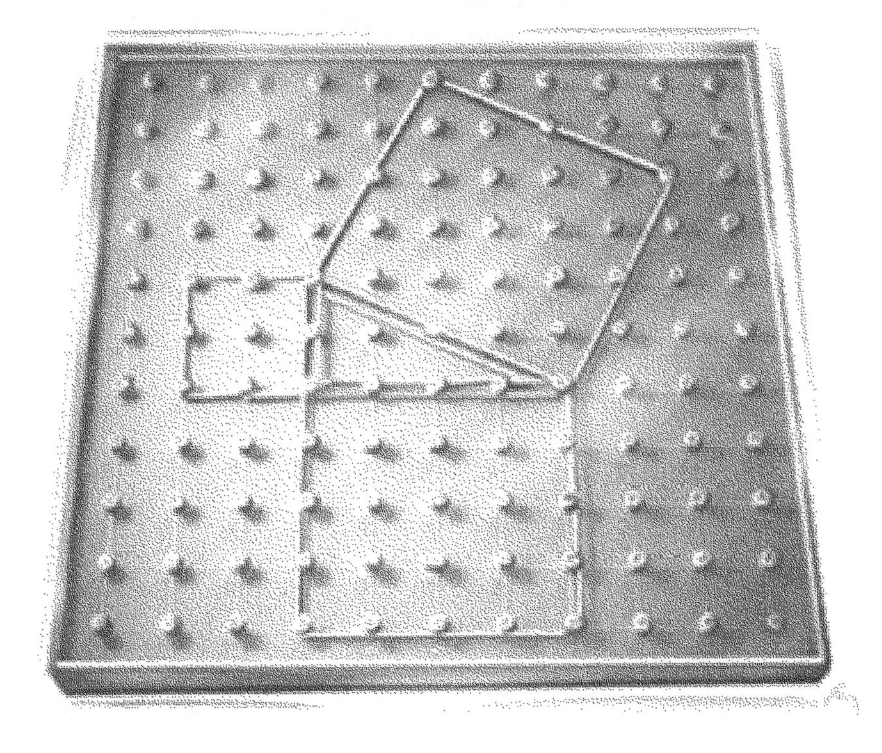

This chapter, which must be preceded by Lab 8.5 (Geoboard Squares), is built around a hands-on approach to the Pythagorean theorem.

- Mathematically, this approach builds the ideas of square and square root on their geometric interpretation; this improves students' grasp of these operations and of the theorem.
- Pedagogically, since this approach is founded on student-discovered techniques for finding area on the geoboard, it is more likely to be meaningful.

Altogether, this approach provides a powerful complement to the traditional one. In general, I have found it preferable to start with the labs found in this chapter before attempting the traditional geometric proofs of the Pythagorean theorem (although Lab 8.5 provides an excellent preview of those). However, the geoboard work is so different in flavor from the traditional curriculum that either order of presentation may work equally well.

In the course of these lessons, I have found the exploration of taxicab geometry to be useful in a variety of ways.

- First, it helps confront directly the misconceptions that many students already have about distance on a Cartesian grid. By discussing this explicitly, it is easier to clarify what is meant by Euclidean distance.
- Second, studying taxi-circles, taxi-ellipses, taxi-equidistance, and so on helps to throw light on the Euclidean equivalents of these concepts.
- Finally, taxicab geometry provides an example of a non-Euclidean geometry that is accessible to all students and helps convey the creativity and playfulness of pure mathematics.

Of course, it is not necessary to master taxicab geometry beyond the very basics, and this is definitely an "enrichment" topic.

See page 214 for teacher notes to this section.

LAB 9.1

Name(s) ______________________

Taxicab Versus Euclidean Distance

Equipment: Geoboard, graph or dot paper

If you can travel only horizontally or vertically (like a taxicab in a city where all streets run North-South and East-West), the distance you have to travel to get from the origin to the point (2, 3) is 5. This is called the taxicab distance between (0, 0) and (2, 3). If, on the other hand, you can go from the origin to (2, 3) in a straight line, the distance you travel is called the Euclidean distance, or just the distance.

Finding taxicab distance: Taxicab distance can be measured between any two points, whether on a street or not. For example, the taxicab distance from (1.2, 3.4) to (9.9, 9.9) is the sum of 8.7 (the horizontal component) and 6.5 (the vertical component), for a total of 15.2.

1. What is the taxicab distance from (2, 3) to the following points?
   a. (7, 9)
   b. (–3, 8)
   c. (2, –1)
   d. (6, 5.4)
   e. (–1.24, 3)
   f. (–1.24, 5.4)

Finding Euclidean distance: There are various ways to calculate Euclidean distance. Here is one method that is based on the sides and areas of squares.

Since the area of the square at right is 13 (why?), the side of the square—and therefore the Euclidean distance from, say, the origin to the point (2,3)—must be $\sqrt{13}$, or approximately 3.606 units.

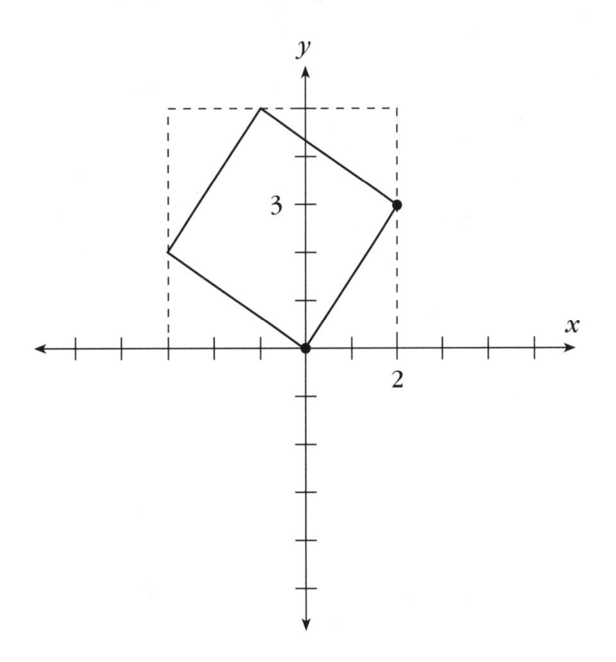

Taxicab Versus Euclidean Distance (continued)

2. What is the Euclidean distance from (2, 3) to the following points?
   a. (9, 7)
   b. (4, 8)
   c. (2, 5.5)
   d. (6, 0)
   e. (1.1, 3)
3. Find as many geoboard pegs as possible that are at a distance 5 from (5, 5). Record your findings on graph or dot paper.
   a. Using taxicab distance
   b. Using Euclidean distance

Discussion

A. Find a formula for the taxicab distance between two points $P_1(x_1, y_1)$ and $P_2(x_2, y_2)$. Call the distance $T(P_1, P_2)$. (**Hint:** Start by figuring out a formula for the case where the points are on a common horizontal or vertical line. The formula should work whether P_1 or P_2 is named first.)

B. In Euclidean geometry, for three points A, B, and C, we always have $AB + BC \geq AC$. This is called the *triangle inequality*. Does this work in taxicab geometry? In other words, do we have $T(A, B) + T(B, C) \geq T(A, C)$? If so, in what cases do we have equality?

C. Which is usually greater, taxicab or Euclidean distance? Can they be equal? If so, in what cases?

D. Explain why the answers to Problem 3a are located on what may be called a *taxi-circle*.

LAB 9.2
The Pythagorean Theorem

Name(s) ______________________

Equipment: Geoboards, dot paper

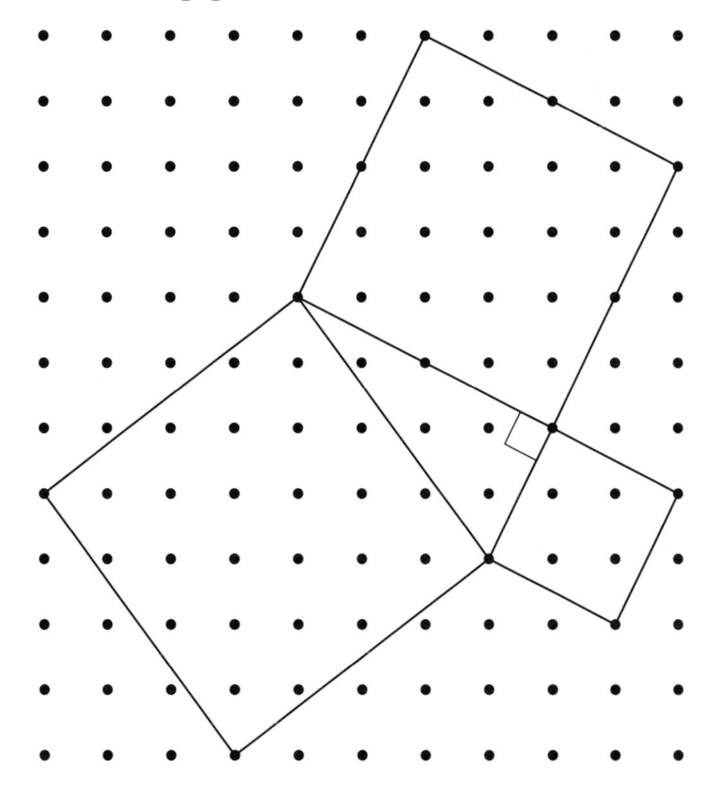

1. The figure above shows a right triangle with a square on each side. Find the areas of the squares.
2. Make your own right triangles on geoboards or dot paper, and draw the squares on the sides, as in the figure. Then, working with your neighbors, fill out the table at right. (Note: The "small" and "medium" squares can be the same size.)
3. Describe the pattern of the numbers in the table. It is called the Pythagorean theorem.
4. State the Pythagorean theorem by completing this sentence: "In a right triangle . . . "

Area of squares		
Small	**Medium**	**Large**

Euclidean distance can be calculated between any two points, even if their coordinates are not whole numbers. One method that sometimes works is to draw a slope triangle (i.e., a right triangle with horizontal and vertical legs), using the two points as the endpoints of the hypotenuse. Then use the Pythagorean theorem.

5. What is the Euclidean distance from (2, 3) to the following points?

a. (7, 9)

b. (−3, 8)

c. (2, −1)

d. (6, 5.4)

e. (−1.24, 3)

f. (−1.24, 5.4)

6. Imagine a circle drawn on dot paper with the center on a dot. How many dots does the circle go through if it has the following radii?

a. 5

b. 10

c. $\sqrt{50}$

d. $\sqrt{65}$

e. $\sqrt{85}$

Discussion

A. Repeat Problem 2 with each of the following. Does the Pythagorean theorem work in these cases? If it fails, how?

a. An acute triangle

b. An obtuse triangle

B. In what parts of Problem 5 did you *not* use the Pythagorean theorem? How did you find those distances?

C. Find as many geoboard isosceles triangles as possible whose legs share a vertex at the origin and whose bases are *not* horizontal, vertical, or at a 45° angle. (Use Euclidean distance for this puzzle. Limit yourself to examples that can be found on an 11 × 11 geoboard. Problem 6 provides a hint.) Record your findings on graph or dot paper.

LAB 9.3
Simplifying Radicals

Name(s) ______________________

Equipment: Geoboard, dot paper

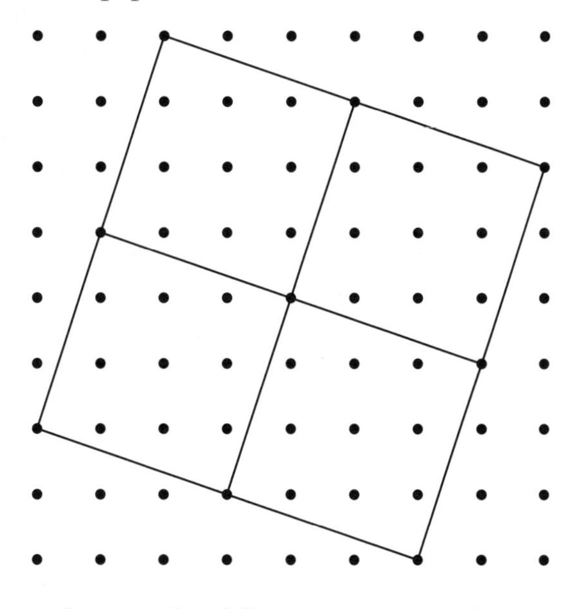

1. In the above figure, what are the following measures?
   a. The area of one of the small squares
   b. The side of one of the small squares
   c. The area of the large square
   d. The side of the large square
2. Explain, using the answers to Problem 1, why $\sqrt{40} = 2\sqrt{10}$.
3. On the geoboard or dot paper, create a figure to show that $\sqrt{8} = 2\sqrt{2}$, $\sqrt{18} = 3\sqrt{2}$, $\sqrt{32} = 4\sqrt{2}$, and $\sqrt{50} = 5\sqrt{2}$.
4. Repeat Problem 3 for $\sqrt{20} = 2\sqrt{5}$ and so on.

 Name(s) ____________

In the figure on the previous page, and in the figures you made in Problems 3 and 4, a larger square is divided up into *a square number of squares.* This is the basic idea for writing square roots in *simple radical form.* The figure need not be made on dot paper. For example, consider $\sqrt{147}$. Since $147 = 3 \cdot 49$, and since 49 is a square number, we can divide a square of area 147 into 49 squares, each of area 3:

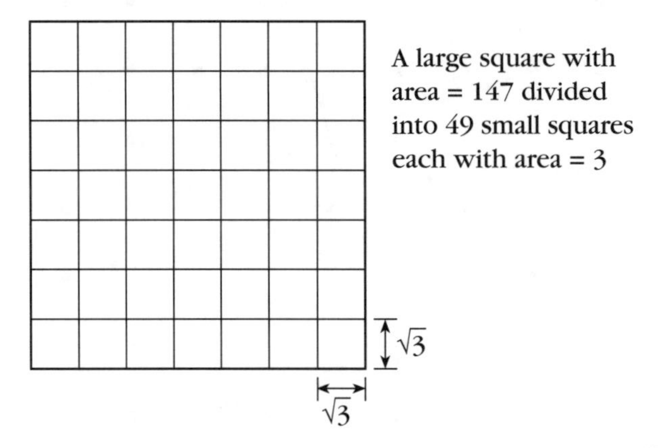

A large square with area = 147 divided into 49 small squares each with area = 3
√3
√3

If you pay attention to the sides of the figure, you will see that $\sqrt{147} = 7\sqrt{3}$. Of course, drawing the figure is not necessary.

5. Write the following in simple radical form.

a. $\sqrt{12}$

b. $\sqrt{45}$

c. $\sqrt{24}$

d. $\sqrt{32}$

e. $\sqrt{75}$

f. $\sqrt{98}$

Discussion

A. Draw a figure that illustrates $4\sqrt{5}$ as the square root of a number.

B. Explain how to use a number's greatest square factor to write the square root of that number in simple radical form. Explain how this relates to the figure above.

LAB 9.4
Distance from the Origin

Name(s) ______________________

Equipment: Geoboard, dot paper

1. What is the distance from each geoboard peg to the origin? Write your answers in simple radical form on the figure below.

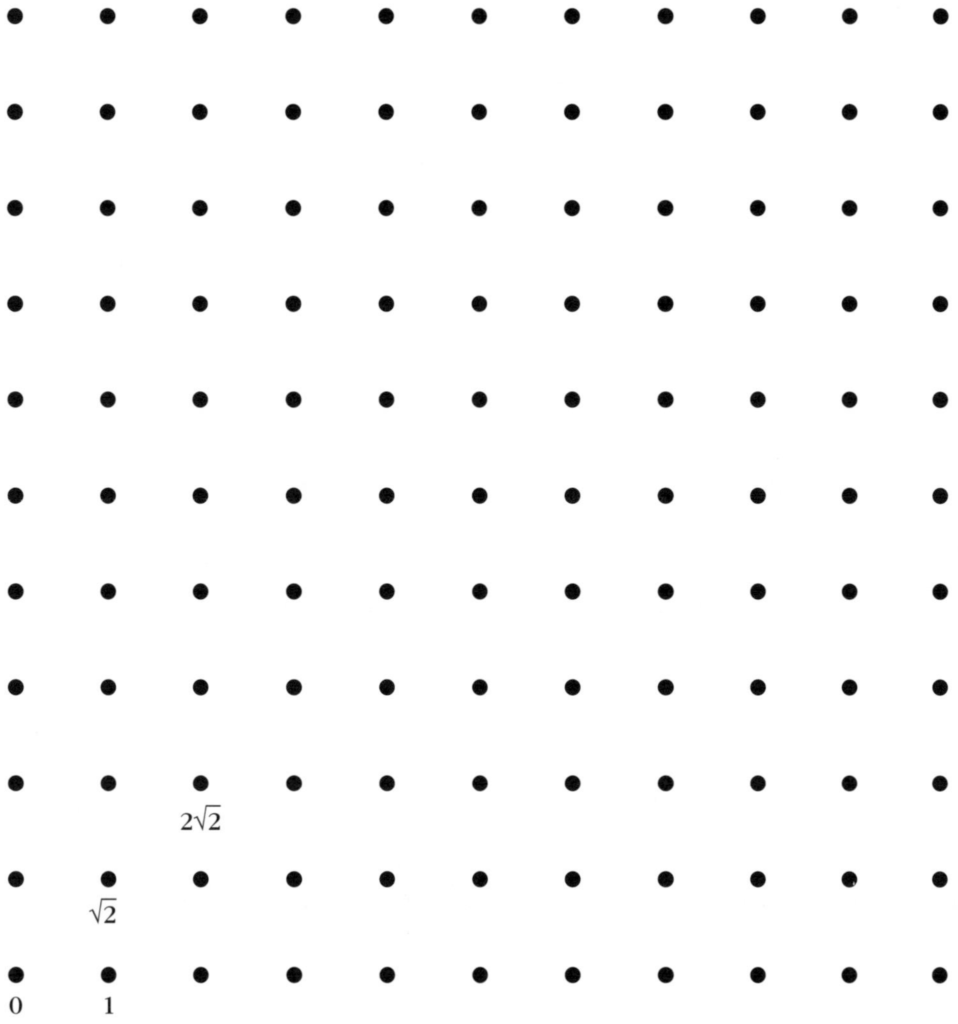

$2\sqrt{2}$

$\sqrt{2}$

0 1

Discussion

A. Discuss any patterns you notice in the distances. Use color to highlight them on the figure. In particular, refer to the following features.

a. Symmetry

b. Slope

LAB 9.5

Name(s) ____________________

Area Problems and Puzzles

■ **Equipment:** Pattern blocks, 1-inch graph paper, assorted dot papers

Problems 1–3 are about the circle geoboard.

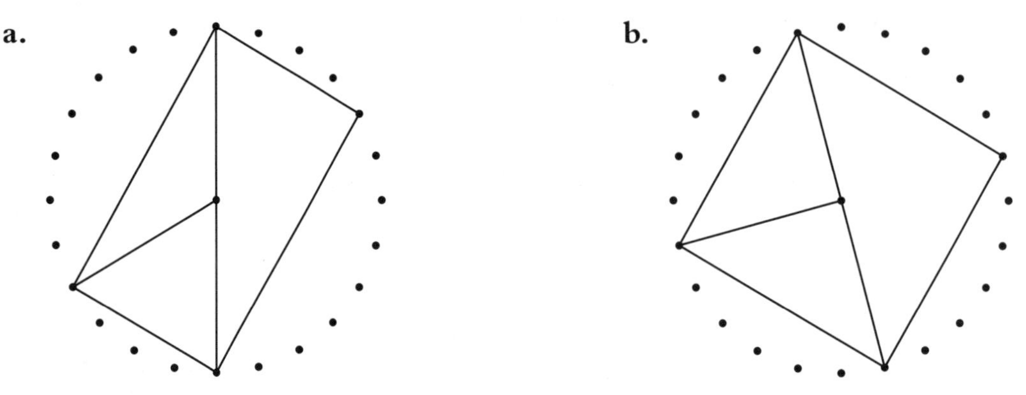

1. Find all the angles in the two figures above.
2. Assuming the radius of the circle is 10, find the lengths of all the segments in the two figures.
3. Find the areas of all the triangles in the two figures.

Problems 4 and 5 are about pattern blocks.

4. Find two ways to cover the figure below with pattern blocks.

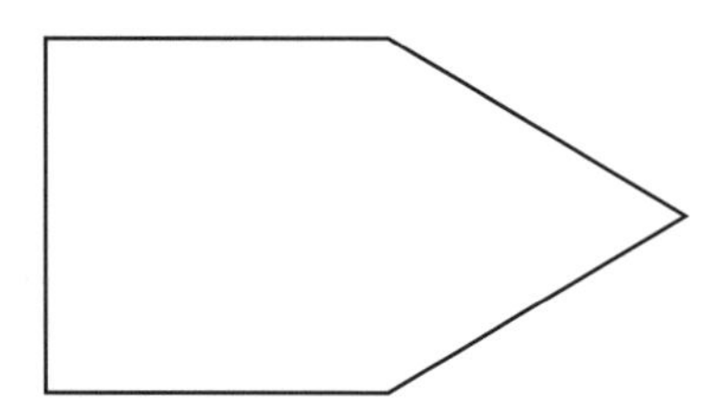

5. Find the area in square inches of each of the six pattern blocks.

The remaining two problems are substantial and are suitable for long-term projects. The first problem is a puzzle with polyominoes.

6. Draw a pentomino shape (five squares joined edge to edge) on 1-inch graph paper. Figure out how to cut it with as few straight cuts as possible so that the pieces can be reassembled to form a square.

The next problem is about generalizing Pick's formula.

7. Is there a formula similar to Pick's formula for area on a geoboard where the pegs are arranged differently than in a square array? (You may research triangular, hexagonal, and rectangular arrays.)

LAB 9.6

Name(s) ____________________

Taxicab Geometry

■ **Equipment:** Dot or graph paper

These problems are substantial. Do not try to do all of them in a short time!

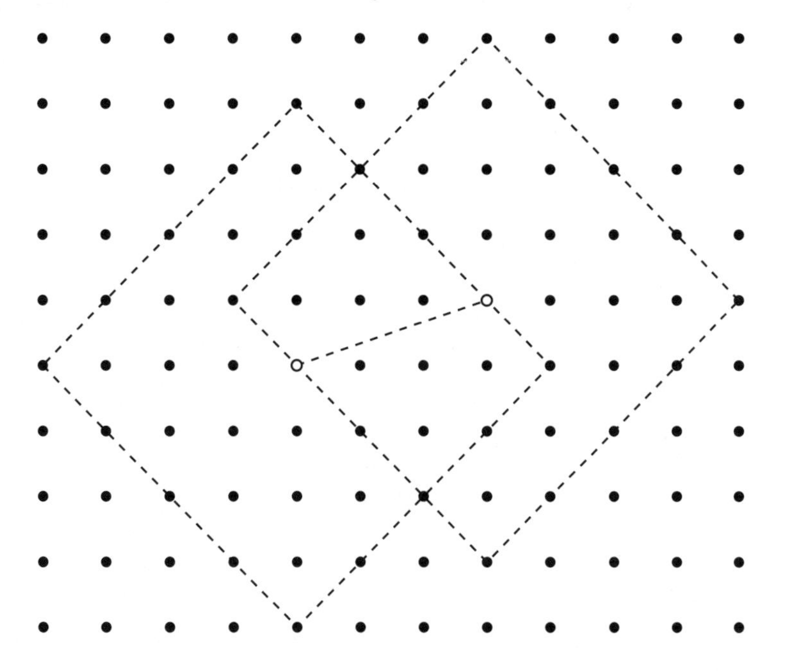

1. If three points A, B, and C are plotted so that, in taxicab distances, $AB = BC = CA$, the points are said to form a *taxi-equilateral* triangle. Sketch a taxi-equilateral triangle that is
   - **a.** Isosceles in Euclidean geometry (**Hint:** The figure above may help.)
   - **b.** Scalene in Euclidean geometry

Problems 2–5 involve taxicab distances to two points.

2. Sketch the set of points that are equidistant from two points A and B in the following cases.
   - **a.** $A(0, 0)$ and $B(6, 0)$
   - **b.** $A(0, 0)$ and $B(4, 2)$
   - **c.** $A(0, 0)$ and $B(3, 3)$
   - **d.** $A(0, 0)$ and $B(1, 5)$
3. Find the set of points P such that $PA + PB = 6$ in the cases listed in Problem 2.
4. Find the set of points P such that $PA + PB = 10$ in the cases listed in Problem 2.
5. Find the set of points P such that $PA/PB = 2$ in the cases listed in Problem 2. (Careful! This last case is particularly tricky.)

Problems 6–9 are about taxi-circles.

6. The number π is the ratio of the perimeter of a circle to its diameter. In Euclidean geometry, $\pi = 3.14159 \ldots$. Find the value of taxi-π.

7. Given the line l with equation $y = 3x$ and the point $P(3, 5)$, construct a taxi-circle with radius 3 that passes through P and touches l in exactly one point. (There are two such circles.)

8. Explain how to find the center of a taxi-circle that goes through two points A and B and, once you have the center, how to sketch the circle. Give examples based on the cases listed in Problem 2.

9. In Euclidean geometry, three noncollinear points determine a unique circle, while three collinear points determine no circle. In taxicab geometry, the situation is somewhat more complicated. Explore different cases, and try to find out when three points determine no circle, one circle, or more than one circle.

Discussion

A. Given two points A and B, how would you find a third vertex P such that ΔABP is taxi-isosceles? (**Hint:** There are two cases: $PA = PB$ and $PA = AB$. And, of course, A and B can be positioned in various ways, such as the ones listed in Problem 2.)

B. Given two points A and B, find the set of points P such that $PA + PB$ is minimal. Investigate the same question for three or even more points.

10 Similarity and Scaling

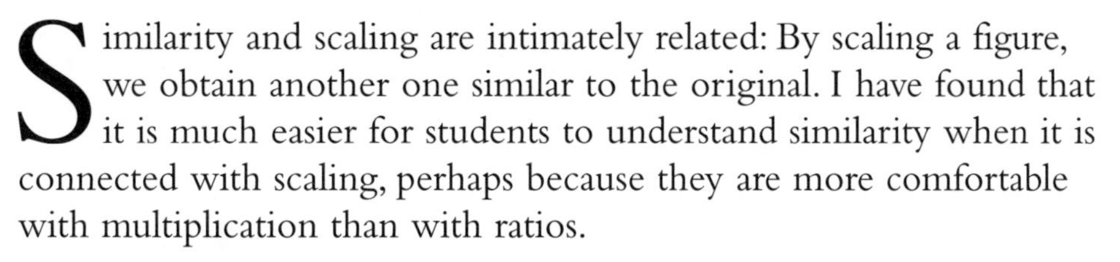

Similarity and scaling are intimately related: By scaling a figure, we obtain another one similar to the original. I have found that it is much easier for students to understand similarity when it is connected with scaling, perhaps because they are more comfortable with multiplication than with ratios.

Two other features of the approach in this section help students get a handle on similarity.

- The work on a Cartesian grid helps students visualize what happens when a figure is scaled.
- The use of the geoboard lattice or the whole numbers of squares in polyominoes helps by providing simpler numbers to work with initially.

In Lab 10.2 (Similar Rectangles), I make a connection between similarity and slope. It is an unfortunate consequence of the artificial boundaries between our algebra and geometry courses that such connections are not made more often.

The other central idea of this section is the relationship between the ratio of similarity and the ratio of areas. This idea is very difficult, as it stands at the intersection of two profound concepts: proportionality and dimension. It certainly cannot be grasped from a brief explanation. In my experience, the most effective way to get it across to students is to explicitly separate out the two dimensions, which we do in Lab 10.1 (Scaling on the Geoboard) and Lab 10.3 (Polyomino Blowups).

See page 222 for teacher notes to this section.

LAB 10.1

Name(s) ____________________

Scaling on the Geoboard

■ **Equipment:** 11 × 11 geoboard, dot paper

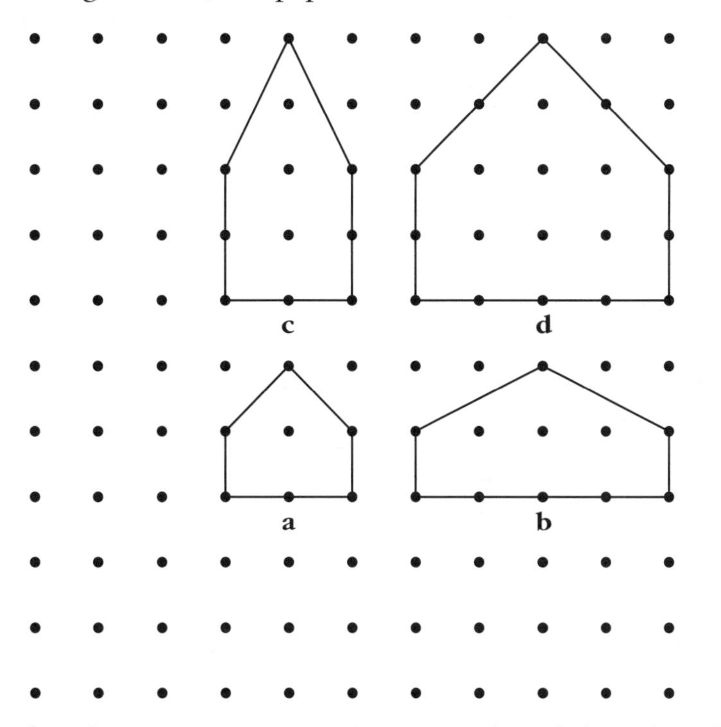

The figure shows four houses. House **a** is the original, and the others have been copied from it following very exact rules.

1. What is the rule that was used for each copy?

2. Among the three copies, two are distorted, and one is a scaled image of the original.

a. Which one is scaled?

b. How would you describe the distortions?

When two figures are scaled images of each other, they are said to be *similar.* Similar figures have *equal angles* and *proportional sides.* The sides of one figure can be obtained by multiplying the sides of the other by one number called the *scaling factor.*

3. What is the scaling factor that relates the two similar houses in the figure above?

4. Make a geoboard triangle such that the midpoint of each side is on a peg. Make your triangle as different as possible from those of your neighbors. Join the midpoints with another rubber band, making a smaller triangle.

   a. Describe the resulting figure, paying attention to equal segments and angles, congruent triangles, parallel lines, and similar figures.

   b. Find the slopes of the lines you believe to be parallel, the length of the segments you believe to be equal, and the scaling factor for the figures you believe to be similar.

5. Make a geoboard quadrilateral such that the midpoint of each side is on a peg. Make your quadrilateral as different as possible from those of your neighbors. Stretch another rubber band around consecutive midpoints, making a smaller quadrilateral.

   a. Describe the resulting figure, paying attention to equal segments and angles and parallel lines.

   b. Find the slopes of the lines you believe to be parallel and the length of the segments you believe to be equal.

Discussion

A. The rules discussed in Problem 1 can be expressed in terms of coordinates (assuming the origin is at the bottom left peg) or in terms of measurements. Explain.

B. For any pair of similar figures, there are two scaling factors relating them. How are the factors related to each other?

C. Try to make a triangle so that two of its sides have midpoints on pegs, but the third does not. What happens? Explain.

D. Problems 4 and 5 are special cases of general theorems that apply to the midpoints of *any* triangle or quadrilateral. State the theorems. Explain how the result in Problem 5 is a consequence of the result in Problem 4.

LAB 10.2
Similar Rectangles

Name(s) ______________________

■ **Equipment:** 11 × 11 geoboard

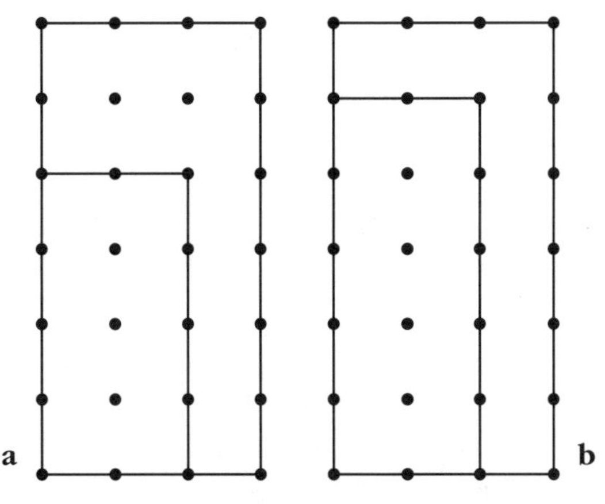

The figure above shows geoboard rectangles nested inside each other.

1. Explain why the two rectangles on the left are similar but the two rectangles on the right are not.

2. Draw the diagonal from the bottom left to the top right of the larger rectangle in **a**. Note that it goes through the top right vertex of the smaller rectangle. Repeat on **b**. How is it different?

3. The rectangles in **a** are part of a set of similar geoboard rectangles. If we include only rectangles with a vertex at the origin (the bottom left peg of the geoboard), the set includes ten rectangles (five horizontal ones and five vertical ones—two of the vertical ones are shown in the figure).

   a. List the rectangles in that set by listing the coordinates of their top right vertex.

   b. What is the slope of the diagonal through the origin for the vertical rectangles?

   c. What is the slope of the diagonal for the horizontal rectangles?

4. Including the set you listed in Problem 3, there are ten sets of similar rectangles (including squares) on the geoboard. Working with your neighbors, find them all, and list all the rectangles in each set. (Again, assume a vertex at the origin, and use the coordinates of the top right vertex for the list.)

5. Working with your neighbors, find every geoboard slope between 1 and 2. (Express the slopes both as fractions and as decimals.) Counting 1 and 2, there are seventeen different slopes.

Discussion

A. Problem 2 is an example of using the *diagonal test for similar rectangles*. Explain.

B. In Problems 3 and 4, how does symmetry facilitate the job of listing the rectangles?

C. What are the advantages of fractional versus decimal notation in Problem 5?

D. Explain how to use the answers to Problem 5 to create lists of geoboard slopes in the following ranges.

- **a.** Between 0.5 and 1
- **b.** Between –1 and –2
- **c.** Between –0.5 and –1

LAB 10.3

Name(s) ______________________

Polyomino Blowups

■ **Equipment:** Interlocking cubes, 1-Centimeter Grid Paper, Polyomino Names Reference Sheet

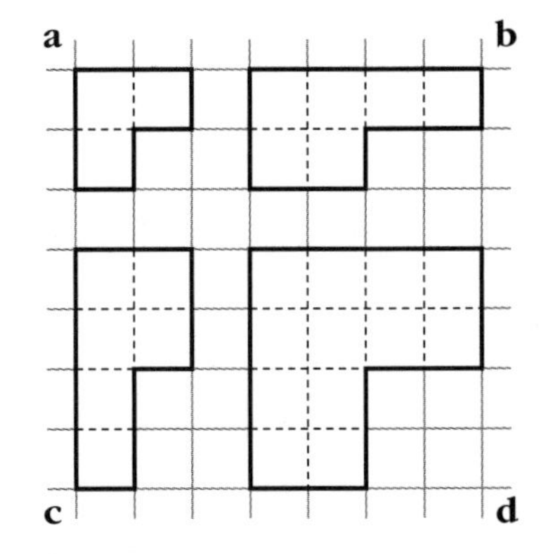

The figure above shows a bent tromino (**a**) and three copies of it. The **b** copy has been doubled *horizontally,* the **c** copy has been doubled *vertically,* and the **d** copy has been doubled in *both* dimensions.

1. Which of the doubled figures is similar to the original? Explain.

2. For each polyomino with area less than 5, draw the polyomino and three copies on grid paper following the pattern in the figure above. As you work, record the perimeter and area of each stretched polyomino in the table.

	Perimeter				**Area**			
		Doubled				**Doubled**		
	Original	**Horiz.**	**Vertic.**	**Both**	**Original**	**Horiz.**	**Vertic.**	**Both**
Monomino								
Domino								
Bent								
Straight								
Square								
l								
i								
n								
t								

3. Study the table for patterns relating the numbers in the different columns. Write down what you observe.

4. The most important patterns are the ones relating the original figure with the similar figure. State the patterns for perimeter and area.

5. Write down a prediction about what may happen to perimeter and area if you triple, and quadruple, a polyomino in both dimensions.

6. Test your prediction by drawing a few polyominoes and their tripled and quadrupled copies. Write down your conclusions.

The following problems are puzzles involving similar polyominoes.

7. **a.** With your interlocking cubes, make tiles in the shape of the **l** and **t** tetrominoes.

   b. On grid paper, draw the five doubled tetrominoes (doubled in both dimensions!).

   c. Use your tiles to cover each doubled figure. Record your solutions.

8. Repeat Problem 7 with the following shapes, being sure to multiply the dimensions both horizontally and vertically.

   a. Tripled tetrominoes, using **l** and **t** tiles

   b. Doubled pentominoes, using P and N tiles

   c. Tripled pentominoes, using P and L tiles

Discussion

A. What is the relationship between the scaling factor and the ratio of perimeters?

B. What is the relationship between the scaling factor and the ratio of areas? Why is this answer different from the answer to Question A?

C. How many tetrominoes does it take to tile a tripled tetromino? How many pentominoes does it take to tile a tripled pentomino? Explain.

D. How many polyominoes does it take to tile a polyomino whose area has been multiplied by k? Explain.

LAB 10.4 Name(s) ____________
Rep-Tiles

Equipment: Interlocking cubes, 1-Centimeter Grid Paper, Polyomino Names Reference Sheet, template

A shape is a rep-tile if it can be used to tile a scaled copy of itself. For example, the bent tromino is a rep-tile, as you can see in the figure below.

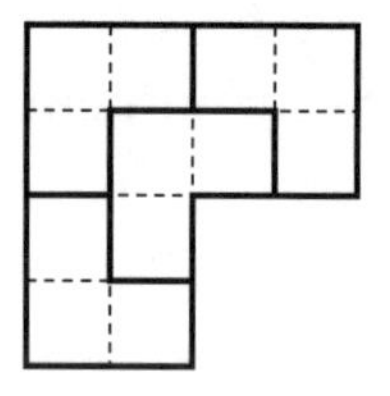

1. The **I** tetromino and the **P** pentomino are rep-tiles. Use cubes and grid paper to tile the doubled and tripled versions of the shapes with the original. Keep a record of your solutions.
2. How many original shapes did you need to tile the following?
   a. The doubled figures
   b. The tripled figures
3. Which other polyominoes of area less than or equal to 5 are rep-tiles? Working with your neighbors, find the tilings to support your answer.
4. Classify the pattern blocks into two groups: those that are rep-tiles and those that are not. (Use the template to make your drawings.)
5. Using each triangle on the template, show that it is a rep-tile.
6. In most cases, it takes four copies of a triangle to show it is a rep-tile. Illustrate the following exceptions.
   a. Show that two right isosceles triangles tile a larger right isosceles triangle.
   b. Show that three half-equilateral triangles tile a larger half-equilateral triangle.
7. Classify the quadrilaterals on your template into two groups: those that are rep-tiles and those that are not.

Discussion

A. When a figure is scaled by a factor k, its area is multiplied by ____. Explain.

B. Prove that *any* triangle is a rep-tile.

C. What are the scaling factors in Problem 6?

D. Find a triangle that can tile a scaled copy of itself using five tiles.

LAB 10.5
3-D Blowups

Name(s) ______________________

■ **Equipment:** Interlocking cubes

In three dimensions, similar solids are obtained by scaling in all three dimensions. The figure below shows an example, with the scaling happening in three steps: width, depth, and height.

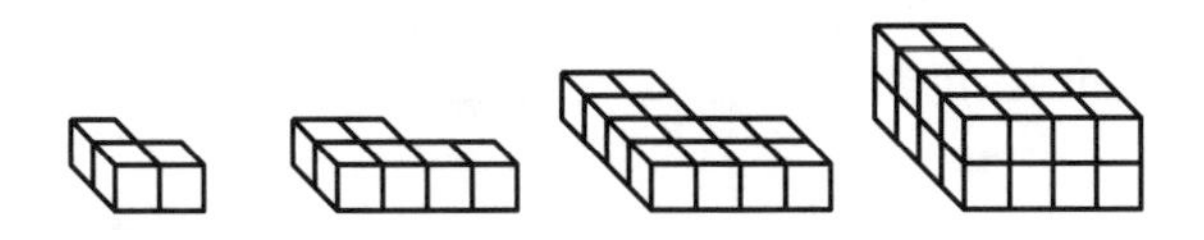

1. The following questions refer to the figure above.
   a. Which two solids are similar?
   b. What is the scaling factor?
   c. What are their surface areas?
   d. What is the ratio of the surface areas?
   e. What are their volumes?
   f. What is the ratio of the volumes?
2. Working with your neighbors, make other pairs of similar solids with your interlocking cubes using a scaling factor of 2, and fill out the table below. Use only a few cubes in the smaller solid!

Surface area			Volume		
Original	Blowup	Ratio	Original	Blowup	Ratio

Name(s) ____________

3-D Blowups (continued)

3. Repeat Problem 2 with a scaling factor of 3. Use even fewer cubes in the smaller solid!

Surface area			Volume		
Original	Blowup	Ratio	Original	Blowup	Ratio

4. Choose one of the solids in the figure below, different from your neighbors' choices. Without using physical cubes, find the surface area and volume for blowups of your solid, and fill out the table. You may use sketches.

Scaling factor	Surface area	Area ratio	Volume	Volume ratio
1				
2				
3				
4				
5				
6				

Discussion

A. When you use a scaling factor k, what happens to each individual unit cube? How does this determine the ratio of surface areas? The ratio of volumes? Explain.

B. Write a formula for the surface area A of a scaled solid as a function of the original surface area A_0 and the scaling factor k. Repeat for the volume.

C. Write a formula for the volume V of the blown-up versions of the solid you studied in Problem 4 as a function of the surface area A and the scaling factor k. Compare your formula with those of your neighbors.

LAB 10.6

Name(s) ____________________

Tangram Similarity

■ **Equipment:** Tangrams

In this lab, write your answers in simple radical form.

1. Fill out this table of tangram perimeters. Part of the first row, where the leg of the small triangle has been taken for the unit of length, has been filled out for you. In the next row, the leg of the medium triangle is the unit, and so on.

Legs			**Perimeters**				
Sm Δ	**Md Δ**	**Lg Δ**	**Sm Δ**	**Md Δ**	**Lg Δ**	**Square**	**Parallelogram**
1	$\sqrt{2}$		$2 + \sqrt{2}$	$2 + 2\sqrt{2}$		4	
	1						
		1					

2. Fill out this table of tangram areas.

Legs			**Areas**				
Sm Δ	**Md Δ**	**Lg Δ**	**Sm Δ**	**Md Δ**	**Lg Δ**	**Square**	**Parallelogram**
1	$\sqrt{2}$		$\frac{1}{2}$	1		1	
	1						
		1					

3. The tangram triangles in the tables above are similar. For each pair of triangles, there are two scaling factors (smaller to larger and larger to smaller). Find all such scaling factors.

4. For each answer in Problem 3, there is a corresponding ratio of areas. What is it?

Name(s) ______________________

5. The figure below shows three similar tangram figures. Check that the scaling factor from smallest to largest is 2. What are the other scaling factors?

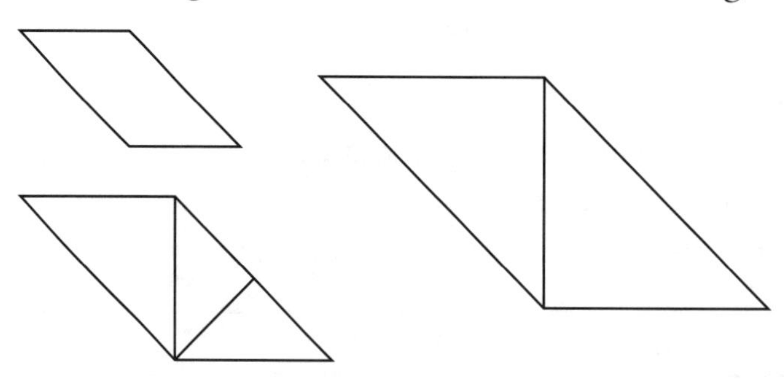

6. **Puzzle:** Find pairs of similar tangram figures using the tangrams from one set only. For each pair, sketch your solution on a separate sheet and record the corresponding scaling factor. Note: Congruent pairs count as similar.

Discussion

A. In the case of tangrams, is it easier to find the ratio of areas or the ratio of similarity of two similar figures? Why?

B. What are some shortcuts for the process of filling out the tables in Problems 1 and 2?

C. Congruent figures have equal angles and proportional sides, so they're also similar. What's the ratio of similarity in congruent figures?

D. In Problem 6, is it possible to find a pair of similar tangram figures that uses all seven pieces?

LAB 10.7
Famous Right Triangles

Name(s) ____________________

	Leg_1	Leg_2	Hypotenuse
a.	1	1	
b.	1	2	
c.	1		2
d.	3	4	
e.		12	13

1. Fill out the table above. If there are square roots, use simple radical form.

Refer to the completed table to solve Problems 2–6.

2. Which triangle is a half-square? Sketch it, indicating the sides and angles.

3. Which triangle is a half-equilateral? Sketch it, indicating the sides and angles.

4. Which triangles have whole-number sides? In such cases, the three numbers are called a *Pythagorean triple.*

5. Find the sides of triangles similar to these five with a scaling factor of 2.

6. Find the sides of triangles similar to these five with a scaling factor of x.

Name(s) ____________________

Famous Right Triangles (continued)

Solve Problems 7–12 two ways.

a. By the Pythagorean theorem

b. By similar triangles

7. A right triangle with sides of length 10 and 20 is similar to one of the five triangles. What are the possibilities for the third side?

8. A square has side 60. How long is the diagonal?

9. A square has diagonal 30. How long is the side?

10. An equilateral triangle has side 70. What is the height?

11. An equilateral triangle has height 40. How long is the side?

12. A right triangle is also isosceles. One of its sides is 50. What are the possibilities for the other sides?

Discussion

A. Consider pattern blocks, tangrams, and the 11×11 geoboard. Which famous right triangle is most relevant to each?

11 Angles and Ratios

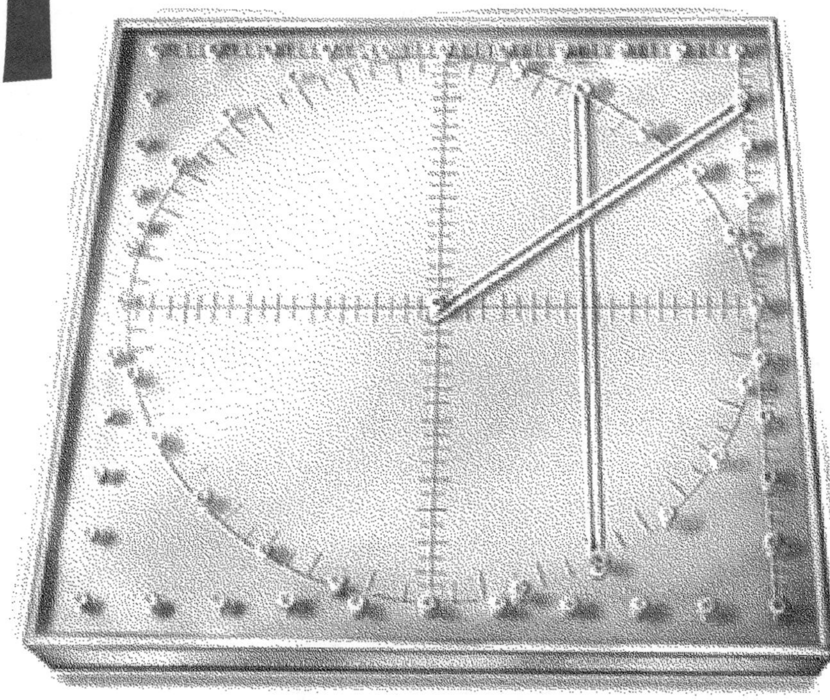

This section provides a new entry into trigonometry. You can use it as a preview or introduction to trig in grades 8–10 or as a complement to a traditional or electronic-based approach in grades 10–11. Of course, the section is far from offering a comprehensive coverage of trigonometry, but it does introduce the basics of right triangle trig, plus the unit circle, in a way that should be accessible to all students.

The essential tool is the 10-cm-radius circle geoboard, or paper versions of it (see p. 245). The design of the board includes:

- a built-in protractor;
- accurately marked x- and y-axes as well as tangent and cotangent lines;
- strategically placed pegs to support an introduction to right triangle and circle trigonometry.

To demonstrate uses of the CircleTrig geoboard on an overhead projector, make a transparency of the paper version.

The section starts with an introduction to the tangent ratio, which is more accessible than sine and cosine and is related to the already familiar concept of slope. The sine and cosine ratios follow. (At the Urban School, we introduce the tangent ratio in the first semester of our Math 2 course and the sine and cosine in the second. This way students' energy can be focused on understanding the underlying concepts, rather than on trying to memorize which of the ratios is which.) Students build tables of those functions and their inverses with the help of the circle geoboard and then use the tables to solve "real world" problems. The names *tangent, sine,* and *cosine* are not used right away, mostly as a way to keep the focus on the geometric and mathematical content. Of course, these names need to be introduced sooner or later, and you will have to use your judgment about whether to do that earlier than is suggested here. If you are using these labs in conjunction with other trig lessons, you may not have the option of postponing the introduction of the names, given that students will recognize the ratios.

In any case, when you do introduce the names, you will need to address the students' question, "Why are we doing this when we can get the answer on the calculator?" The answer that has worked for me is, "You can use your calculator if you want, but for these first few labs, show me that you understand the geometric basis of these ideas." The calculator is the method of choice if you're looking for efficiency, but the geometric representation in the circle is a powerful mental image, which will help students understand and remember the meaning of the basic trig functions.

See page 231 for teacher notes to this section.

LAB 11.1 Angles and Slopes

Name(s) ______________________

■ **Equipment:** CircleTrig geoboard, CircleTrig geoboard sheet

The CircleTrig geoboard and the CircleTrig geoboard sheet include ruler and protractor markings.

1. Label the rulers on the sheet (not on the actual geoboard) in 1-cm increments.
2. Label the protractor markings on the sheet in 15° increments. Start at 0° on the positive x-axis, going counterclockwise. Include angles greater than 180°.
3. Repeat Problem 2, going clockwise. In this direction, the angles are considered negative, so this time around, 345° will be labeled –15°.

Think of the line with equation $y = mx$, and of the angle θ it makes with the positive x-axis (θ is the Greek letter *theta*). Each slope m corresponds to a certain angle between –90° and 90°.

You can think about this relationship by making a right triangle on your CircleTrig geoboard like one of those shown below. The legs give the rise and run for the slope of the hypotenuse. You can read off the angle where the hypotenuse crosses the protractor to find the angle that corresponds to a given slope.

Note that even though both examples at right show positive slopes, you can use the CircleTrig geoboard to find the angles corresponding to negative slopes as well.

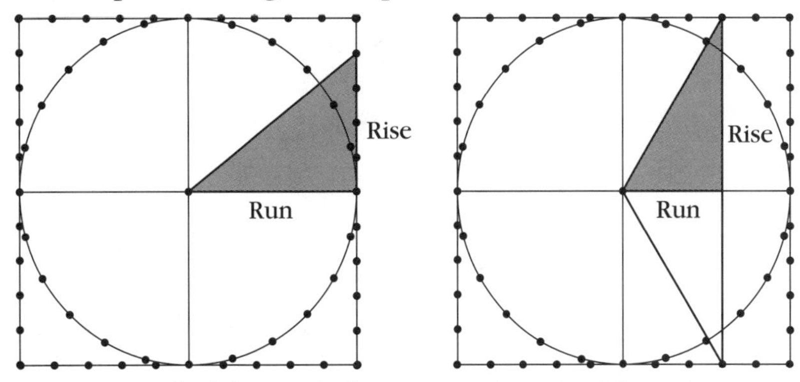

Two ways to find the angle for a given slope (Problem 4)

You can also find slopes that correspond to given angles. Two examples of how to do this are shown at right. In the first example, you can read rise off the y-axis and run off the x-axis. In the second example, the rubber band is pulled around the 30° peg, past the right edge. You can read rise off the ruler on the right edge. (What would run be in that example?)

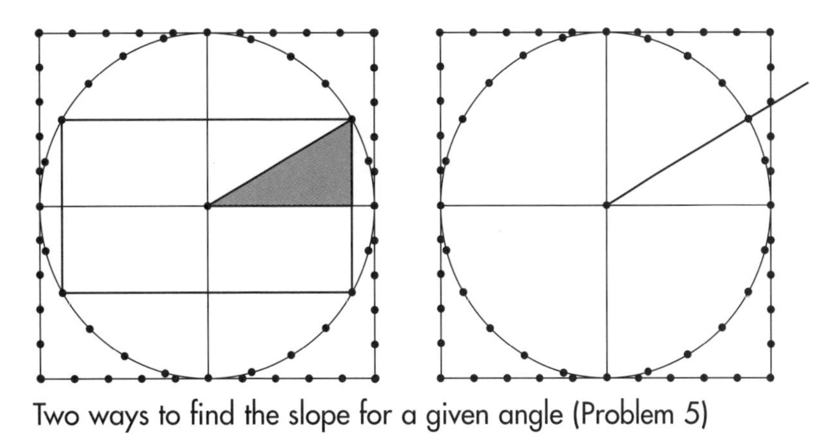
Two ways to find the slope for a given angle (Problem 5)

4. Fill out the table below. Continue a pattern of going around the outer pegs of the geoboard to supply slopes where the table is blank. For angles, give answers between –90° and 90°. (That is, make your slope triangles in the first and fourth quadrants.)

m	0	0.2	0.4	0.6	0.8	1	1.25	1.67				–5		
θ											90°			

5. Fill out the table below.

θ	0°	15°	30°	45°	60°	75°	90°	105°	120°	135°	150°	165°	180°
m													

Discussion

A. What patterns do you notice when filling out the tables? What is the relationship between the slopes of complementary angles? For what angles is the slope positive? Negative? 0? For what angles is the slope between 0 and 1? Greater than 1?

B. Why is there no slope for the angle of 90°?

C. Explain how you chose one or another of the four types of slope triangles to help you fill out the tables.

D. Some of the slope triangles you used to fill out the tables are "famous right triangles." Check that the angles and slopes you found are correct by comparing your answers with those you got in Lab 10.7.

LAB 11.2 Using Slope Angles

Name(s) ______________________

■ **Equipment:** The tables from Lab 11.1, CircleTrig geoboard, CircleTrig geoboard paper

1. How tall is the flagpole?

2. How far is the boat from the edge of the cliff?

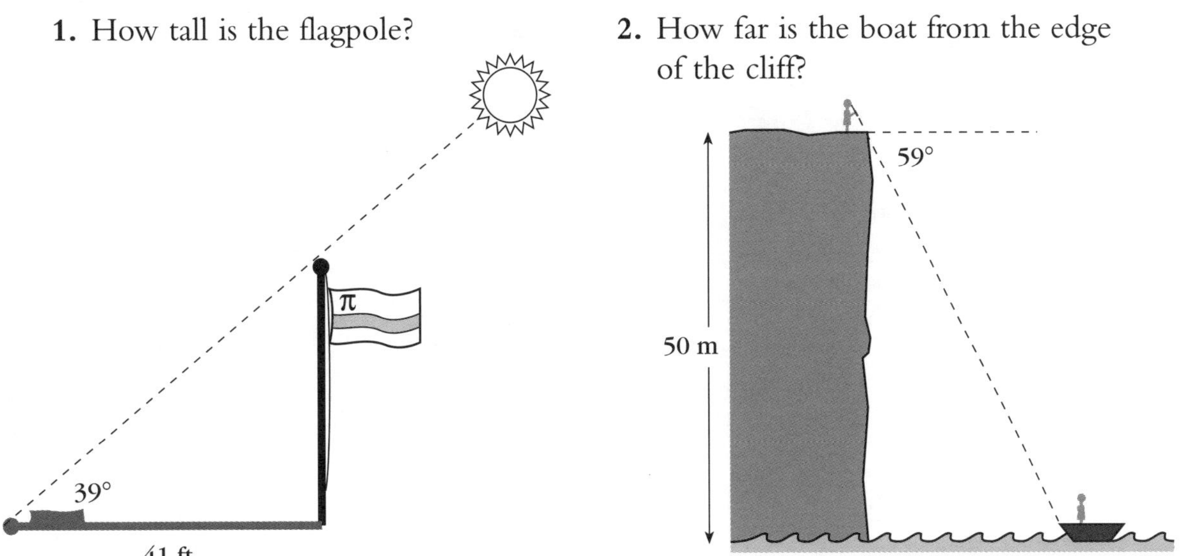

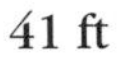

For the remaining problems, make your own sketches on a separate sheet of paper.

3. Looking down at a boat from a 30-m-high lighthouse, an observer measures an angle of 15° below the horizontal.

a. Sketch this.

b. How far is the boat from the base of the lighthouse?

4. A ski lift rises 200 meters for a run of 250 meters. What angle does it make with the horizontal?

5. At a certain time of day, a 33-ft flagpole casts a 55-ft shadow. What is the angle made by the sun's rays with the horizontal?

6. The banister of a straight staircase makes an angle of 39° with the horizontal. The stairs connect two floors that are 10 feet apart.

a. How much horizontal space does the staircase take?

b. If steps are 8 inches high, how wide are they?

7. You stand on a cliff, looking down at a town in the distance. Using a map, you find that the town is 1.2 km away. The angle your line of vision makes with the horizontal is 11°. How high is the cliff?

8. A right triangle has a 15° angle and a short leg of 18 units. How long is the long leg?

9. A right triangle has a 75° angle and a short leg of 18 units. How long is the long leg?

Name(s) ______________________

Discussion

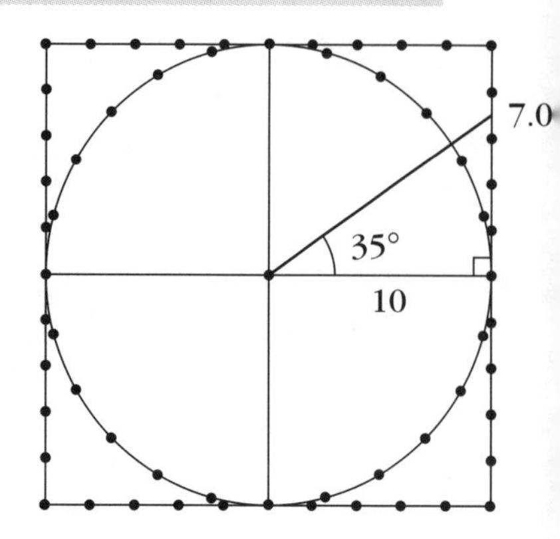

A. The problems in this lab have been rigged to use only angles and slopes that you included in the tables. The figure suggests a method for finding the slopes and angles in other cases by using a rubber band *between* pegs on the CircleTrig geoboard or a ruler on the paper geoboard. Explain the technique.

B. Find the angles for slopes of 10, 25, 100. What happens as the slope gets bigger and bigger?

LAB 11.3

Name(s) ______________________

Solving Right Triangles

Equipment: CircleTrig geoboard or CircleTrig geoboard paper

Right triangles are crucial in many parts of math, physics, and engineering. The parts of a right triangle are (not counting the right angle):

- two legs;
- one hypotenuse;
- two acute angles.

You will soon know how to find all the parts given a minimum amount of information. Finding all the parts is called *solving* a triangle.

For each problem below, answer the question and work the example. Use the figure to keep track of what you know.

1. Given one acute angle, what other parts can you find? (Example: One acute angle is 21°.)

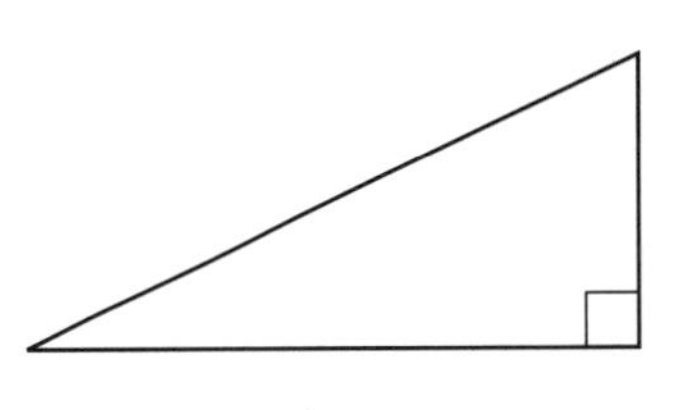

2. Given one side, what other parts can you find? (Example: One side is 3.)

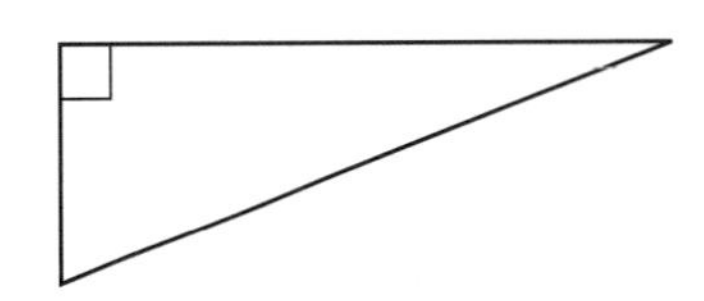

3. Given two legs, what other parts can you find? (Example: One leg is 4 and the other is 5.)

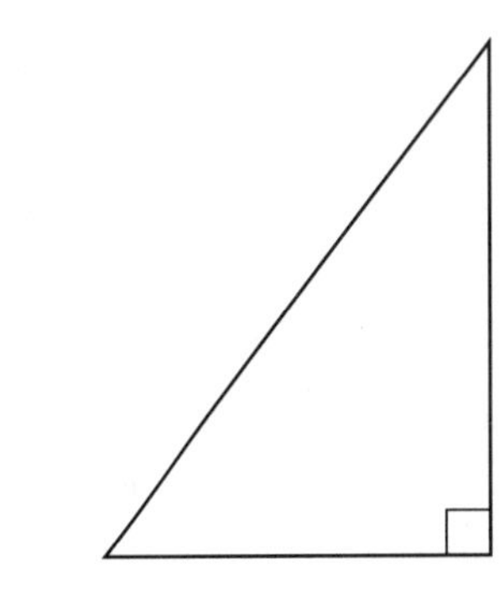

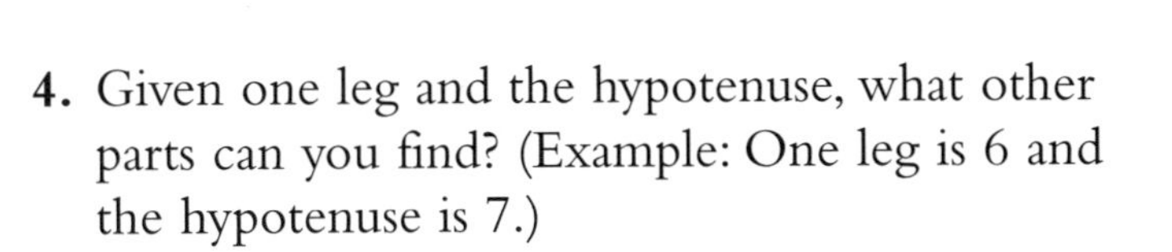

4. Given one leg and the hypotenuse, what other parts can you find? (Example: One leg is 6 and the hypotenuse is 7.)

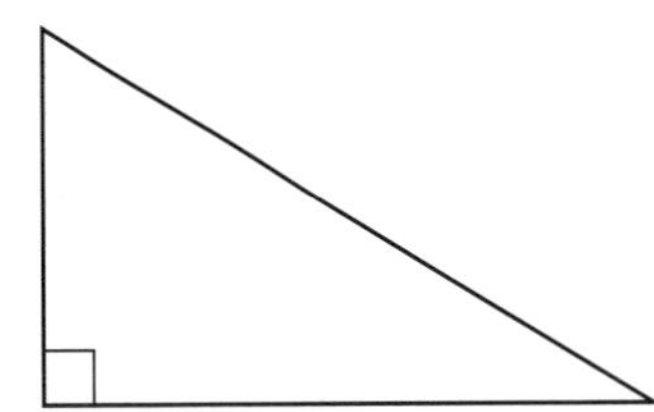

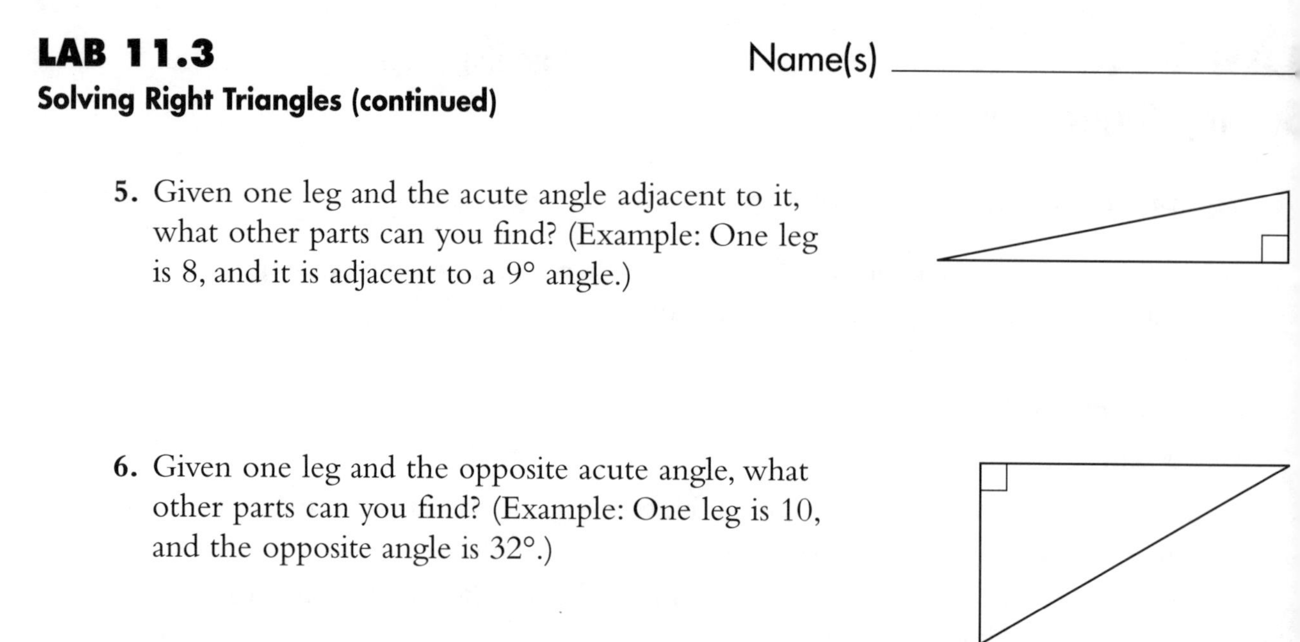

5. Given one leg and the acute angle adjacent to it, what other parts can you find? (Example: One leg is 8, and it is adjacent to a 9° angle.)

6. Given one leg and the opposite acute angle, what other parts can you find? (Example: One leg is 10, and the opposite angle is 32°.)

7. Given the hypotenuse and an acute angle, what other parts can you find? (Example: The hypotenuse is 4, and one angle is 65°. **Hint:** Create your own right triangle with angle 65° and scale it.)

Discussion

A. What is the minimum amount of information necessary to completely solve a right triangle?

B. When you know only one leg, is there a way to know if it is the long leg or the short leg (or whether the legs are the same length)?

C. Problem 7 is much more difficult than the others. Why?

LAB 11.4 Name(s) ______________________
Ratios Involving the Hypotenuse

- **Equipment:** CircleTrig geoboard, CircleTrig geoboard paper

In Labs 11.1 and 11.2, we were working with three numbers: the two legs of a right triangle (which we thought of as rise and run) and an angle. Given any two of those, it was possible to find the third. In some right triangle situations, however, the three numbers you have to work with could be the hypotenuse, one leg, and an angle. To address such problems, we will use two ratios involving the hypotenuse:

$$\frac{\text{adjacent leg}}{\text{hypotenuse}} \quad \text{and} \quad \frac{\text{opposite leg}}{\text{hypotenuse}}.$$

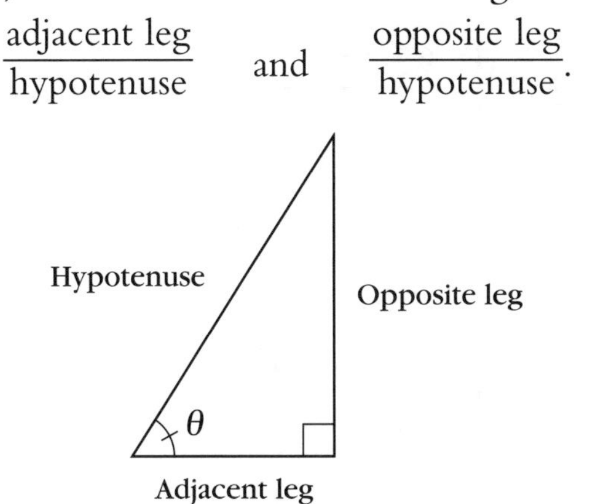

To find those ratios, use your CircleTrig geoboard. To find lengths of adjacent and opposite legs, recall how you found slopes for given angles in Lab 11.1. See the figure at right. Using this method, the length of the hypotenuse will always be the same (what is it?), which makes it easy to write the ratios without using a calculator. Enter your results in the table below. Since you can find lengths to the nearest 0.1 cm, express your ratios to the nearest 0.01 unit.

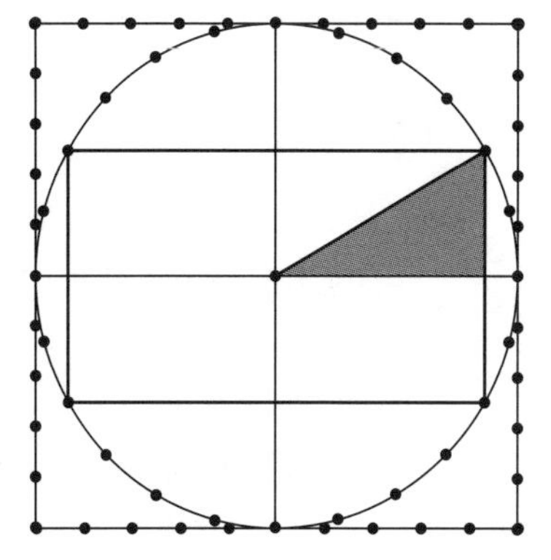

θ	opp/hyp	adj/hyp
0°		
15°		
30°		
45°		
60°		
75°		
90°		

LAB 11.4

Name(s) ____________________

Ratios Involving the Hypotenuse (continued)

The figure at right shows one way to find an angle given an opp/hyp ratio of 0.4. Stretch a rubber band to connect two 4-cm pegs. Read off the angle where the rubber band crosses the protractor markings. (Can you see why the opp/hyp ratio is 0.4 in this example?) You can use a similar method to find angles for given adj/hyp ratios. Enter your results in the tables below.

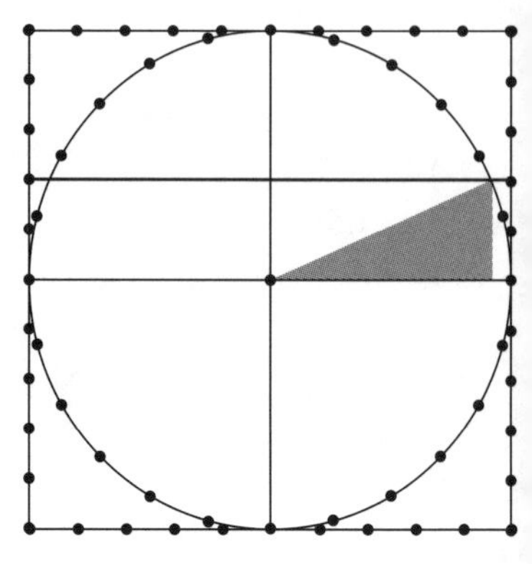

opp/hyp	θ
0	
0.2	
0.4	
0.6	
0.8	
1	

adj/hyp	θ
0	
0.2	
0.4	
0.6	
0.8	
1	

Discussion

A. When filling out the tables, look for patterns. What is the relationship between the ratios for complementary angles? For what angles do we have ratios of 0? 1?

B. Some of the triangles you used to fill out the tables are "famous right triangles." Check that the angles and ratios you found are correct by comparing your answers with those you got in Lab 10.7.

C. Can the $\frac{\text{opp}}{\text{hyp}}$ ratio, or the $\frac{\text{adj}}{\text{hyp}}$ ratio, be greater than 1? Explain.

LAB 11.5 Using the Hypotenuse Ratios

Name(s) ______________________

■ **Equipment:** CircleTrig geoboard

Example: Find the area of a triangle with sides of 5 cm and 7 cm and an angle of 15° between those two sides.

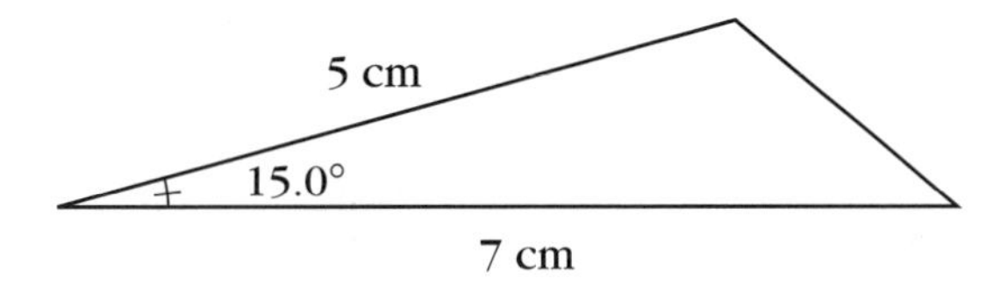

If we use the 7-cm side as the base, we need a height to calculate the area.

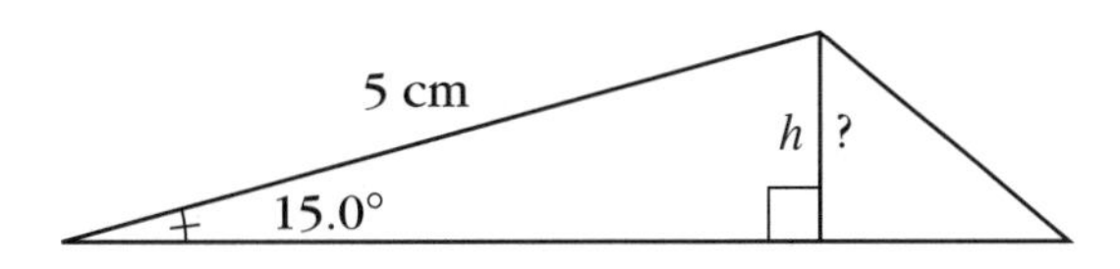

Using the table we made in the previous lab, we see that for a 15° angle the ratio opp/hyp = 0.26.
So we have $\frac{h}{5} = 0.26$.

1. What are the height and area of the triangle described above?

Solve the following problems by a similar method. Always start by drawing a figure.

2. What is the area of a triangle with sides of 4 cm and 6 cm and a 24° angle between those sides?

3. A 12-ft ladder is propped up against a wall. This ladder is safest if it is at a 75° angle from the horizontal. How far should the base of the ladder be from the wall?

4. What are the acute angles of a 3, 4, 5 triangle?

5. What is the acute angle of a parallelogram with sides of 3 cm and 8 cm and area of 14.4 cm^2?

6. A flagpole is held up by wires. You measure the distance from where one of the wires is attached to the ground to the foot of the pole and get 12 m. You measure the angle the wire makes with the horizontal and get 78°. How long are the wires?

Discussion

A. Using the CircleTrig geoboard or the paper CircleTrig geoboard, how would you find the angles and ratios that are between the ones in the table you made in Lab 11.4? The figure shows an example for 35°.

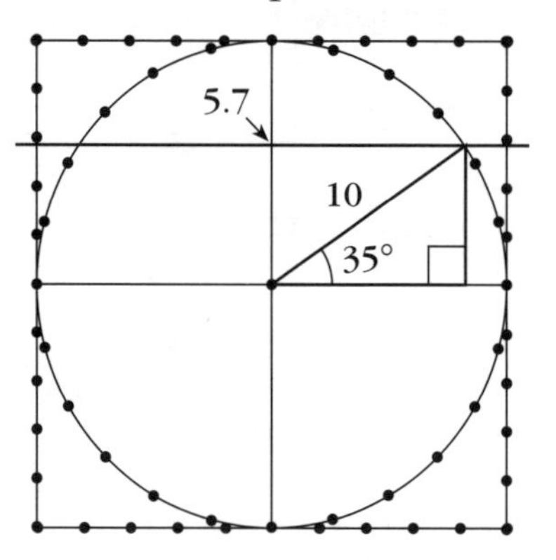

The part of mathematics that studies ratios and angles is called *trigonometry*. Each of the ratios we have used in the past few labs has a name and notation, as shown below.

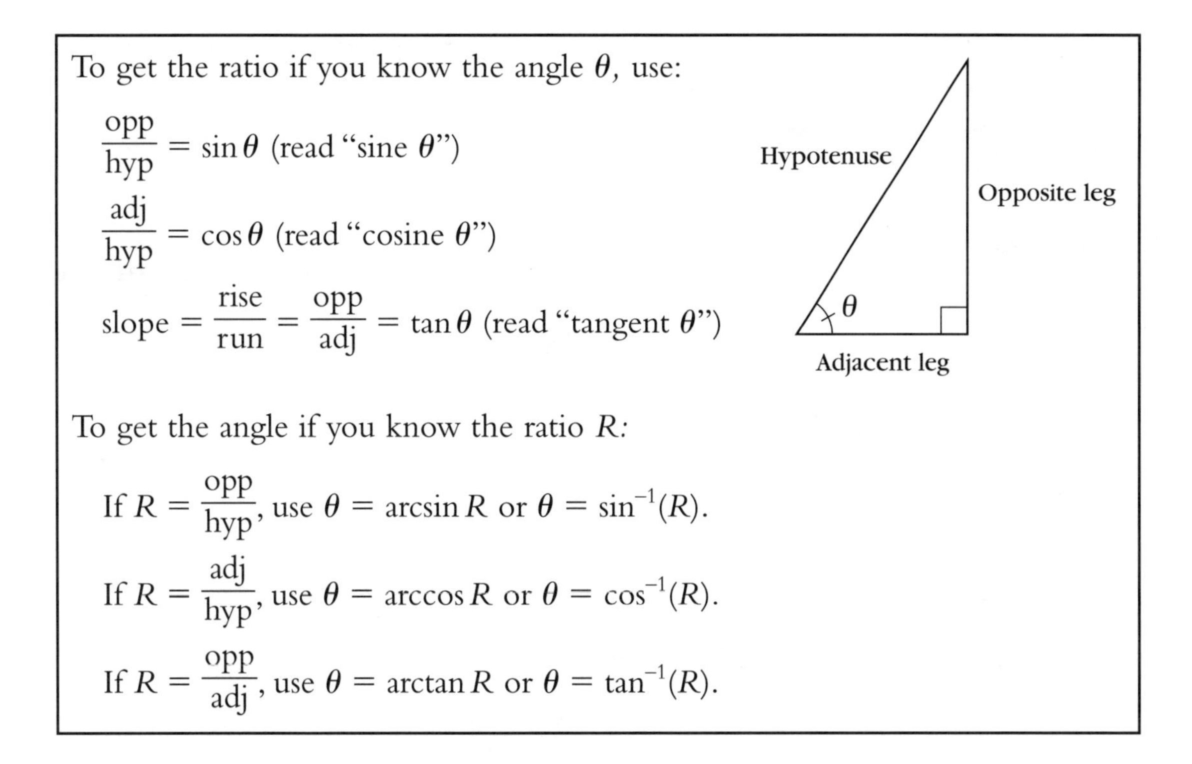

To get the ratio if you know the angle θ, use:

$\frac{\text{opp}}{\text{hyp}} = \sin\theta$ (read "sine θ")

$\frac{\text{adj}}{\text{hyp}} = \cos\theta$ (read "cosine θ")

$\text{slope} = \frac{\text{rise}}{\text{run}} = \frac{\text{opp}}{\text{adj}} = \tan\theta$ (read "tangent θ")

To get the angle if you know the ratio R:

If $R = \frac{\text{opp}}{\text{hyp}}$, use $\theta = \arcsin R$ or $\theta = \sin^{-1}(R)$.

If $R = \frac{\text{adj}}{\text{hyp}}$, use $\theta = \arccos R$ or $\theta = \cos^{-1}(R)$.

If $R = \frac{\text{opp}}{\text{adj}}$, use $\theta = \arctan R$ or $\theta = \tan^{-1}(R)$.

As you may suspect, engineers and physicists do not use geoboards to solve problems of the type we looked at in Lab 11.5. In the old days, they used tables, not unlike the ones we constructed but much more detailed. Later, they also used slide rules. Nowadays, they use calculators and computers. Scientific calculators have been programmed to give you the ratios, given the angles, or the angles, given the ratios. All you need to know are the official names of the three ratios we have been working with. To remember them, you can use one of the following mnemonics:

soh-cah-toa

soppy, cadjy, toad

LAB 11.6
The Unit Circle

Name(s) ______________________

Equipment: CircleTrig geoboard or CircleTrig geoboard paper, Trigonometry Reference Sheet

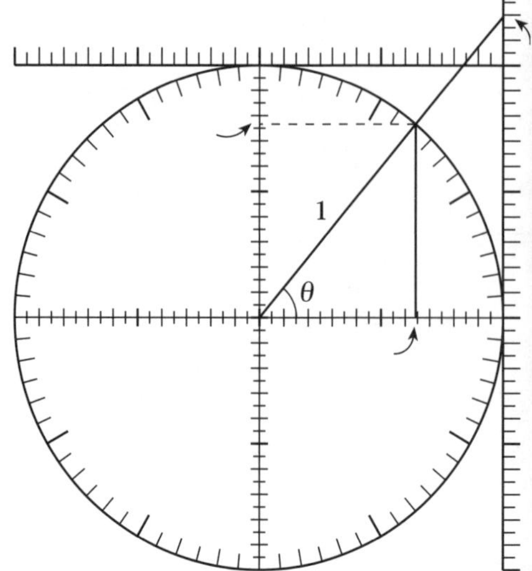

At right is a figure that shows a *unit circle,* in other words, a circle with radius 1. Angles on the unit circle are measured from the x-axis, with counterclockwise as the positive direction. Angles can be any positive or negative number.

1. Explain why on the unit circle all three trig ratios can be thought of as ratios over 1. Use an example in the first quadrant, such as the one in the figure at right.

2. Mark on the figure above where the sine, cosine, and tangent of the angle θ can be read on the axes with no need to divide.

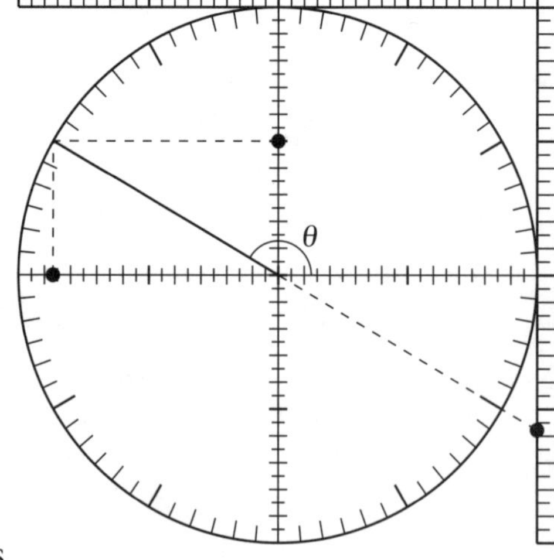

The cosine and sine of an angle can be defined respectively as the x- and y-coordinate of the corresponding point on the unit circle. This definition works for any angle, not just the acute angles found in a right triangle.

As for the tangent, you already know how to find the tangent for any angle: It is the slope of the corresponding line (see Lab 11.1: Angles and Slopes). The *tangent* ratio is so named because it is related to a line tangent to the unit circle: specifically, the vertical line that passes through the point (1, 0). If you extend the side of an angle (in either direction) so that it crosses the tangent line, the tangent of the angle can be defined as the y-coordinate of the point where the angle line and the tangent line cross. Note that the angle line, when extended through the origin, makes two angles with the positive x-axis—one positive and one negative. These two angles have the same tangent since they both define the same line.

3. Mark on the figure above where the sine, cosine, and tangent of the obtuse angle θ can be read on the axes with no need to divide.

4. Show 15° and 165° angles on a unit circle. How are their sines related? Their cosines? Their tangents? Explain, with the help of a sketch.
5. Repeat the previous problem with 15° and 345°.
6. Repeat with 15° and 195°.
7. Repeat with 15° and 75°.
8. Study the figures on the previous page, and find the value of $\sin^2\theta + \cos^2\theta$ (that is, the square of $\sin\theta$ plus the square of $\cos\theta$). Explain how you got the answer.

Discussion

A. Refer to the figure at the beginning of the lab. Explain why you get the same ratios whether you use the smaller or the larger triangle in the figure. For each trig ratio, which triangle is more convenient?

B. Use the smaller triangle to prove that $\tan = \frac{\sin}{\cos}$.

C. For a large acute angle, such as the 75° angle in Problem 7, the tangent can no longer be read on the axis at the right of the unit circle, as its value would lie somewhere off the page. However, the axis that runs along the top of the unit circle is useful in such cases. How is the number you read off the top axis related to the tangent of the angle?

D. Two angles α and β (alpha and beta) add up to 180°. Draw a unit circle sketch that shows these angles, and explain how their trig ratios are related.

E. Repeat Question D with two angles α and β that add up to 90°.

F. Repeat Question D with two angles α and β whose difference is 180°.

G. Repeat Question D with two angles α and β that add up to 360°.

H. Repeat Question D with two angles α and β that are each other's opposite.

I. Explain why the answer to Problem 7 is called the Pythagorean identity.

LAB 11.7 Name(s) ____________

Perimeters and Areas on the CircleTrig™ Geoboard

- **Equipment:** CircleTrig geoboard or CircleTrig geoboard paper, Trigonometry Reference Sheet

Find the perimeter and area of each figure, assuming the circle has a radius of 10 cm.

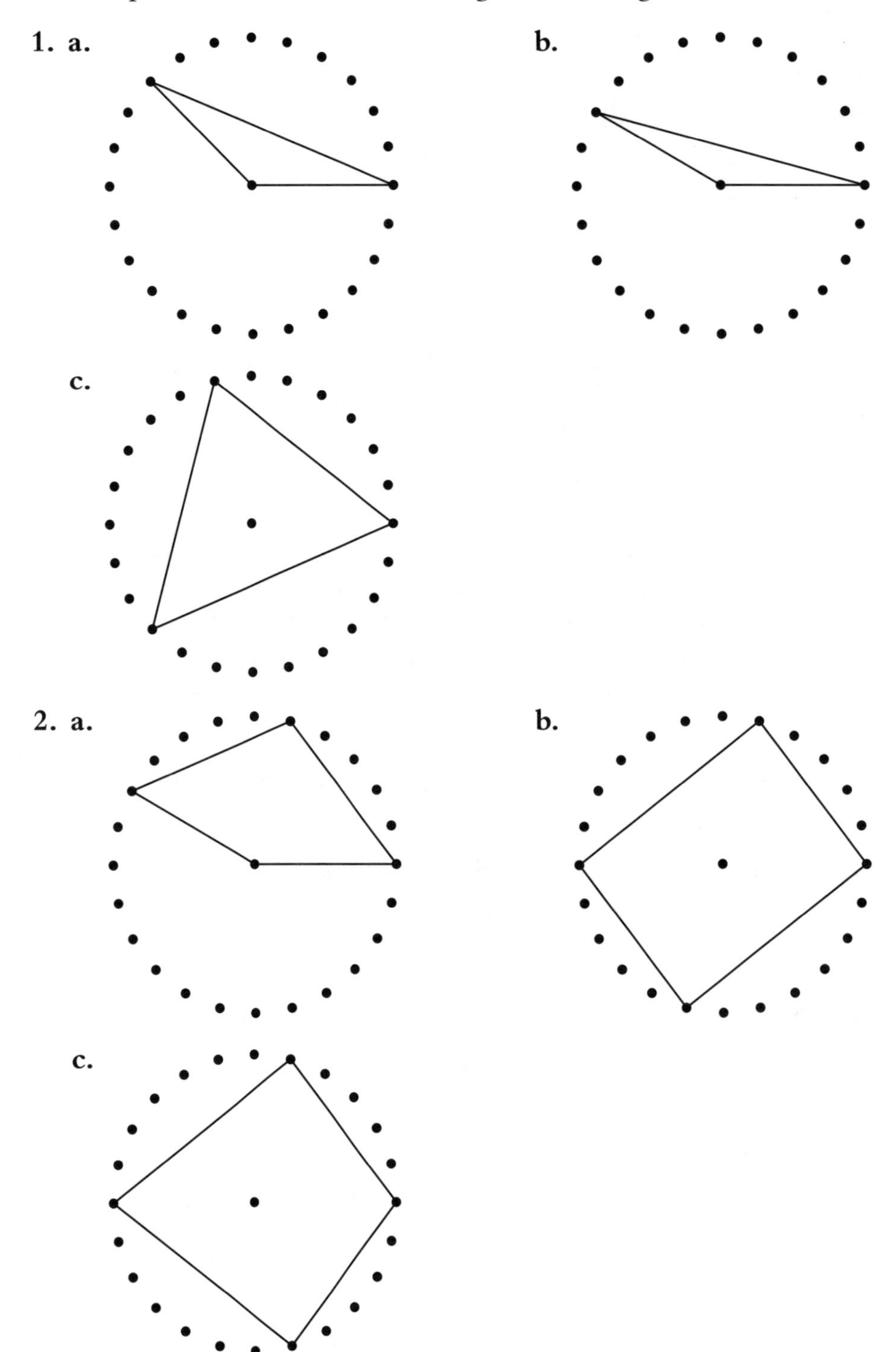

3. Find the perimeter and area for the following regular polygons, inscribed in a circle with a radius of 10 cm.

   a. Equilateral triangle

   b. Square

   c. Hexagon

   d. Octagon

   e. Dodecagon

   f. 24-gon

Discussion

A. The key to all the calculations in this lab is understanding how to find the base and height of an isosceles triangle the legs of which are 10-cm radii. Draw a figure for this situation, and assume a vertex angle of 2θ. What are the formulas for the base and the height? What are the formulas for the perimeter and area of the triangle?

B. If you place a rubber band around the entire circle of the CircleTrig geoboard, you get a regular 24-gon. How close is it to the actual circle in perimeter and area?

LAB 11.8

Name(s) ______________________

"π" for Regular Polygons

You probably know these formulas for a circle of radius r. (P is the perimeter, and A is the area.)

$$P = 2\pi r \qquad A = \pi r^2$$

It follows from these formulas that $\pi = \frac{P}{2r} = \frac{A}{r^2}$.

For the purposes of this lesson, we will define the "radius" of a regular polygon to be the radius of the circle in which the polygon would be inscribed (in other words, the distance from the center of the polygon to a vertex). Furthermore, we will define π_P and π_A for a regular polygon as follows:

$$\pi_P = \frac{P}{2r} \quad \text{and} \quad \pi_A = \frac{A}{r^2}.$$

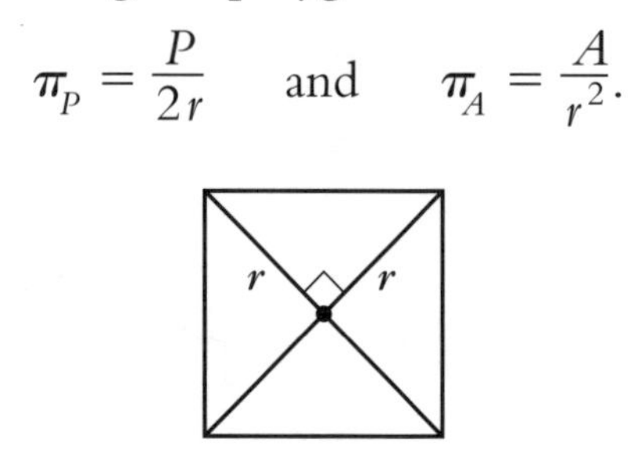

1. For example, consider a square of "radius" r (see the figure above).
   - **a.** What is the perimeter?
   - **b.** What is π_P?
   - **c.** What is the area?
   - **d.** What is π_A?
2. Think of pattern blocks. What is π_P for a hexagon?
3. Use the figure below to find π_A for a dodecagon. **Hint:** Finding the area of the squares bounded by dotted lines should help.

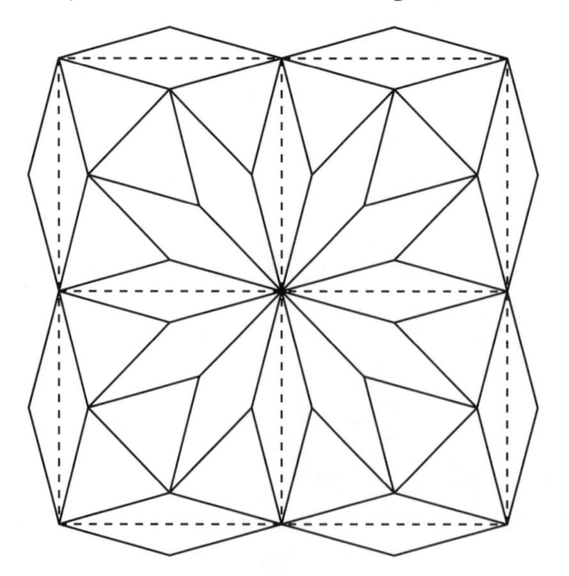

4. In general, you will need trigonometry to calculate π_P and π_A. Do it for a regular pentagon. **Hint:** It is easiest to assume a "radius" of 1; in other words, assume the pentagon is inscribed in a unit circle. Be sure to use a calculator!

5. Fill out the table below. Here, n is the number of sides of the polygon. **Hint:** Start by mapping out a general strategy. Use sketches, and collaborate with your neighbors. You may want to keep track of your intermediate results in a table. Use any legitimate shortcut you can think of.

n	π_P	π_A
3		
4		
5		
6		
8		
9		
10		
12		
16		
18		
20		
24		

Discussion

A. What patterns do you notice in the table?

B. Write a formula for π_P and for π_A, as a function of n.

C. What happens for very large values of n?

D. Of course, these values of π for polygons, while interesting, are not very useful. Why is the real value of π (for circles) more significant?

1 Angles

Lab 1.1: Angles Around a Point

Prerequisites: There are no prerequisites other than knowing that there are 360° around a point. I often use this lab on the first day of school.

Timing: To get the most out of this activity, you should plan to spend at least two periods on it. Students need time to understand what is expected, to search for many solutions, and to discuss their work as they go.

There are many ways to solve the problem with, say, two colors and three blocks. It is enough to find one of the ways to circle that number.

The essential purpose of this lesson is to develop a rough quantitative sense of angle measurement without dwelling on the actual numerical values of the measurements. Make sure that all students are able to think (and therefore talk) about the activity at some level. However, if some students are ready for, and use, actual angle measurements, do not discourage that! Expect all students to get into this more formally in the next lab. When students start thinking that they have found every possible answer, you may reveal that there are 42 solutions altogether. This helps clarify things if there are actually many unfound solutions, but it may sabotage the quality of the discussion, since it takes away the opportunity to discuss Question E, which itself is an incentive to think about Questions F–I. Note that the impossible cases are of two types: logical and geometric. For example, solutions with four colors and three blocks are logically impossible. The cases that are impossible by virtue of the size of angles are more interesting from the point of view of the purpose of the lab.

Answers

Colors:	How many blocks you used:									
all blue	③	④	⑤	⑥						
all green	~~3~~	~~4~~	~~5~~	⑥						
all orange	③	④	~~5~~	~~6~~						
all red	③	④	⑤	⑥						
all tan	~~3~~	④	~~5~~	~~6~~		⑧				⑫
all yellow	③	~~4~~	~~5~~	~~6~~						
two colors	③	④	⑤	⑥	⑦	⑧	⑨	⑩	⑪	~~12~~
three colors	③	④	⑤	⑥	⑦	⑧	⑨	⑩	~~11~~	~~12~~
four colors	③	④	⑤	⑥	⑦	⑧	⑨	~~10~~	~~11~~	~~12~~
five colors	~~3~~	~~4~~	⑤	⑥	⑦	~~8~~	~~9~~	~~10~~	~~11~~	~~12~~
six colors	~~3~~	~~4~~	~~5~~	~~6~~	~~7~~	~~8~~	~~9~~	~~10~~	~~11~~	~~12~~

There are 42 solutions altogether.

Discussion Answers

A. Green, orange, and yellow offer unique solutions because, for each one, all angles are the same size.

B. Using the smaller angle, it takes 12 blocks. Each time you want to add a large angle, you have to remove five small ones, so altogether you have removed four blocks.

C. The blue and red blocks are interchangeable because they have the same angles as each other.

D. Answers will vary.

E. Answers will vary.

F, G, H. See chart. In order to use many blocks, the angles have to be small. But the more colors that are required, the more you are forced to use the blocks that do not have small angles.

I. Even if you use the smallest angles for each block, their sum is still greater than 360°.

Lab 1.2: Angle Measurement

Prerequisites: The previous lab (Angles Around a Point) is recommended.

This activity should help students consolidate their grasp of the concept of angle to the point that they are ready to learn how to use a protractor if they do not yet know how.

If they have no idea about how to solve Problem 2, suggest they arrange blocks around a central point, as in Lab 1.1. Discourage the use of a protractor for this problem, as the point here is to use logic and develop understanding.

You may have the students cut out random paper triangles and find the approximate measures of their angles with the protractor they made in Problem 3.

To use the template protractor for drawing angles in Problem 5, suggest the following procedure:

a. Draw one side of the angle; mark the vertex.

b. Place the center of the protractor on the vertex and line up the desired measurement with the existing side.

c. Draw the other side along the edge of the template.

Answers

1. **a.** 72°

   b. 120°

   c. Answers will vary.

   d. 30°

   e. 90°

2. Red and blue: 60° and 120°; yellow: 120°; orange: 90°; tan: 30° and 150°; green: 60°
3. See student work.
4. All angle sums should be about 180°.
5. See student work.

Lab 1.3: Clock Angles

Prerequisites: The previous lab (Angle Measurement) is recommended.

This exploration is wide open and can serve to assess student understanding. The better that students understand angles (and proportions), the more difficult they can make this challenge for themselves.

The discussion questions can help prepare students for the activity.

Answers

Student reports will vary. The angle at 5 o'clock is 150°; at 5:30, 15°.

A goal for your best students is to find the angle at completely arbitrary times such as 1:23. At that time, the hour hand has moved 23/60 of 30° from the 1, so it makes a 41.5° angle with the 12. The minute hand has moved 23/60 of 360° from the 12, so it makes a 138° angle with the 12. So the angle between the hands is $138 - 41.5 = 96.5°$.

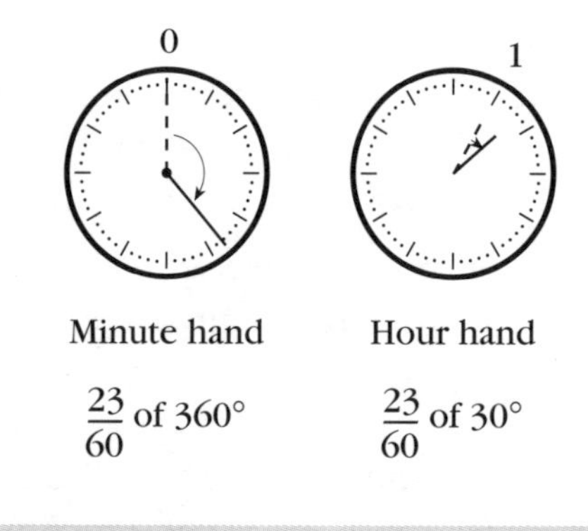

Discussion Answers

A. 30°; 0.5°

B. 360°; 6°

Lab 1.4: Angles of Pattern Block Polygons

Prerequisites: Lab 1.2 (Angle Measurement) is recommended.

The main purpose of this lab is to practice using angle measurements in the context of a broader problem. It also introduces the idea that the sum of the angles in a polygon is a function of the number of sides. Students should express their understanding of this in the summary.

In Problems 3–6, the angles we are adding are the angles of the polygon, not all the angles in the constituent pattern blocks. To clarify this, you may show the figure at right on the overhead.

For this pattern block hexagon, add only the marked angles, excluding the six unmarked pattern block angles. Starting at the bottom left and proceeding clockwise, the sum of the polygon angles is 90 + (90 + 60) + (60 + 60) + 120 + (60 + 90) + 90 = 720°.

If students have trouble with Problems 7–9, give hints and encourage cooperative work. Resist the temptation to reveal the formula prematurely, as this often has the effect of decreasing students' motivation and intellectual engagement. Instead, let students use whatever method they can to solve the problems. If they cannot solve them, tell them that this idea will be reviewed in future lessons. There will be plenty of time to teach the formula then if you think that is important.

Answers

1. Green: 180°; red, blue, tan, orange: 360°; yellow: 720°
2. The four-sided shapes have the same sum.
3. Polygons will vary, but will be six-sided.
4. Polygons will vary, but will be five-sided.
5. Answers will vary.
6. Student sketches will vary. Angle sums in table should be 180°, 360°, 540°, 720°, 900°, 1080°, 1260°, 1440°, 1620°, 1800°.
7. 3240°. Explanations will vary.
8. 25. Explanations will vary.
9. This is impossible because 450 is not a multiple of 180.
10. To get the sum of the angles, subtract 2 from the number of sides and multiply by 180. $S = (n - 2)180$.

Lab 1.5: Angles in a Triangle

This is a "getting ready" activity, which will help prepare students for the following several labs.

Prerequisites:

Definitions:

acute, right, and *obtuse* angles;
equilateral, isosceles, and *scalene* triangles.

Theorems:

The angles of a triangle add up to 180°.
The base angles of an isosceles triangle are equal.
All angles of an equilateral triangle are equal.

If students are familiar with the theorems listed above, you could assign this page as homework.

If students are not familiar with the theorems, you may introduce them informally as described below.

- Have students cut three paper triangles—obtuse, right, and acute—from unlined paper (preferably colored paper).

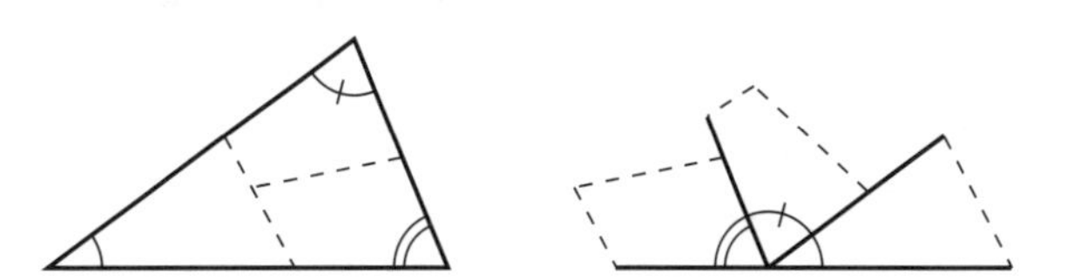

  Then, have them show by tearing and rearranging the angles of each triangle that they add up to 180°. (They can paste the torn and rearranged angles in their notebooks.)

- Hand out a sheet with several triangles on it. Have the students cut out the isosceles triangles on one copy of the sheet and verify that they can superimpose them onto themselves after flipping them over. This verifies that the base angles are equal. Similarly, they can flip and rotate the equilateral triangle in a variety of ways.

You could mention to students that they could find many additional triangles on the template by using two consecutive sides of other polygons and connecting them.

Answers

1. No. There needs to be a third angle, and the sum of two right angles or two obtuse angles is already equal to or greater than 180°.
2. **a.** 60°, 60°, 60°

   b. Answers will vary. The angles should all be less than 90°. Two should be equal.

   c. 45°, 45°, 90°

d. Answers will vary. One angle should be greater than 90°. The other two should be equal.

e. Answers will vary. All angles should be less than 90°. None should be equal.

f. Answers will vary. One angle should be 90°. None should be equal.

g. Answers will vary. One angle should be greater than 90°. None should be equal.

3. 30°, 60°, 90°

4. Right isosceles

5. In an equilateral triangle the angles are all 60°, so they are not right or obtuse.

6. The right triangles: c and f

7. Answers will vary.

8. In a right triangle, the two acute angles *add up to 90°*.

9. See student work.

Lab 1.6: The Exterior Angle Theorem

Prerequisites: Students should know that the sum of the angles in a triangle is 180°.

You can use this activity to preview, introduce, or review the exterior angle theorem. It is a "getting ready" activity, which will help prepare your students for the next lab.

At the very beginning of the activity, make sure students understand the definition of exterior angle. You may use Questions A–C. Another way to get this across, which ties in with some of the labs in Section 3, is to point out that the exterior angle is the turn angle for someone who is walking around a triangle.

Some students may not have the mathematical maturity to really internalize an algebraic understanding of this and will rely instead on a sense of what is happening with the numbers. This is why Problems 4–6 and Questions D and E are important in laying the groundwork for understanding the theorem. Writing explanations for Problems 6c–9 should help cement what understanding is there.

Problem 10 is challenging, though students with a more developed algebraic sense should have no trouble with it. Another approach to it is based on the "walking polygons" approach mentioned above (and detailed in Section 3).

Answers

1. 80°, 100°

2. Answers will vary.

3. Answers will vary, but one of them will always be 100°.

4. a. 115°

b. 65°

5. a. 115°

b. 65°

6. 123° in all cases. Explanations will vary.

7. x°. Explanations will vary.

8. An exterior angle of a triangle is always equal to *the sum of the two nonadjacent interior angles*. Explanations will vary.

9. 90°. Yes. It follows from the fact that the exterior angle at the right angle is also a right angle.

10. a. Always 360°

b. Explanations will vary.

Discussion Answers

A. Three pairs of two or six altogether

B. 180°

C. See student work.

D. Obtuse. In Problems 4 and 5, the exterior angle is acute, so the corresponding interior angle is obtuse. In Problem 6a, $\angle C = 113°$; in 6b, $\angle C = 103°$.

E. x° and $(50 - x)$°; sum: 50°

Lab 1.7: Angles and Triangles in a Circle

Prerequisites:

Definitions:

circle and *radius*.

Theorems:

triangle sum theorem,
isosceles triangle theorem,
exterior angle theorem.

This is an extension that involves a lot of work nd is probably most appropriate in a high chool geometry course, in preparation for the nscribed angle theorem and its proof. However, his lab can be valuable even with less ambitious oals, such as developing a feel for the different ypes of triangles; providing an interesting xperience in using the three theorems listed bove; and discovering Thales's theorem (any ngle inscribed in a half-circle is a right angle). n that case, it is sufficient to stop when most tudents have finished Problem 4, and perhaps Problem 5.

Problems 6 and 7 are mostly designed to prepare tudents for understanding the general proof of he inscribed angle theorem. Instead of doing hem now, you may return to these problems ater, perhaps with only circle geoboard paper ather than actual geoboards. You can even eturn to actual geoboards *after* students know he inscribed angle theorem, as an interesting way to practice applying the theorem.

iming: This lab will take more than one period. Depending on how much discussion time you want to include, it could take up to three periods. t may not be necessary for all students to do he activity to the very end. If you think your tudents are "getting the picture" or are getting ired of this, you may interrupt them at the end of the last period you want to spend on this lab or a wrap-up discussion of what was learned.

Answers

, 2. Because all the radii of the circle are equal, only isosceles and equilateral triangles are possible. The angle at the center of the circle can be found by noticing that the angle between any two consecutive pegs is $360°/24 = 15°$. Since the triangle is isosceles, the other two angles are equal and can be obtained by subtracting the central angle from 180° and dividing by 2.

3, 4. Drawing an additional radius from the center to the third vertex allows you to divide the triangle into two sub-triangles. Then, you can find the angles by using the same method as in Problem 1. Another, faster way to get there is with the exterior angle theorem, which allows you to use the fact that the central angle in one sub-triangle is exterior to the other sub-triangle and therefore equal to twice its noncentral angles.

5. They are all right triangles. Proofs will vary. One way to do it is to observe that the central angles in the two sub-triangles add up to 180°. Since they are equal to twice the noncentral angles, those must add up to 90°.

6, 7. Again, drawing additional radii is a good strategy.

Discussion Answers

A, B, C, D. See solution to Problems 1 and 2.

E. All triangles in Problem 6 are acute; all triangles in Problem 7 are obtuse.

Lab 1.8: The Intercepted Arc

Prerequisites: The previous lab (Angles and Triangles in a Circle) is recommended.

This activity is an extension that provides an additional step in preparing students for the formal proof of the inscribed angle theorem.

Many students find the definitions that open this activity extraordinarily difficult to understand. You may use this activity to preview or reinforce what they will learn from the introduction in the textbook.

Use the discussion questions and the circle geoboard to follow up on this activity. Encourage the students to use the theorem rather than resort to drawing additional radii and using the cumbersome methods from the previous lab.

Answers

1. Inscribed angle; central angle; intercepted arc
2. $\overset{\frown}{AQ}$ for both

3. 25°. Explanations will vary.
4. $c = a/2$. This follows from the exterior angle theorem.
5. $\angle b = 90°$; $\angle c = 25°$; $\angle d = 45°$; $\angle APB = 70°$
6. Answers will vary.
7. $\angle APB = (1/2)\angle AOB$. Explanations will vary.
8. An inscribed angle is equal to half the corresponding central angle.
9. Because of the exterior angle theorem, $\angle c = (1/2)\angle a$ and $\angle d = (1/2)\angle b$. By adding these equations, you get $\angle c + \angle d = (1/2)\angle a + (1/2)\angle b$. Factoring yields $\angle c + \angle d = (1/2)(\angle a + \angle b)$. In other words, $\angle APB = (1/2)\angle AOB$.
10. Figures and proofs will vary. The proofs should be similar to the one in Problem 9, using subtraction instead of addition.

Discussion Answers

A. 52.5°, 75°, 30°, 67.5°

B. Left: 22.5°, 67.5°, 90°. Right: 52.5°, 60°, 67.5°.

C. 120°, 60°, 120°

D. Use arcs of 90°, 120°, and 150°. This will work, since they add up to 360°.

Lab 1.9: Tangents and Inscribed Angles

Prerequisites: The inscribed angle theorem

Two theorems about tangents are previewed (but not proved) here. This lab uses the circle geoboard, with which students are now comfortable, to seed these concepts. At this point, students probably do not need to actually use the physical geoboard, but if any of them want to, do not discourage that.

The best strategy for Problem 4 is to add and subtract known angles in order to find unknown ones.

Answers

1. a. $\overset{\frown}{QA}$, 15°
   b. $\overset{\frown}{QB}$, 37.5°
   c. $\overset{\frown}{QC}$, 67.5°
   d. $\overset{\frown}{QD}$, 82.5°
   e. Explanations will vary.
2. a. $\overset{\frown}{QP}$
   b. 90°
3. Perpendicular
4. Explanations will vary.
   a. $\overset{\frown}{EP}$, 67.5°
   b. $\overset{\frown}{FP}$, 37.5°
   c. $\overset{\frown}{GP}$, 15°
   d. $\overset{\frown}{HGP}$ (or $\overset{\frown}{HFP}$, $\overset{\frown}{HEP}$, $\overset{\frown}{HQP}$), 105°
5. Each angle measurement is indeed half of the intercepted arc, which equals the corresponding central angle.

Lab 1.10: Soccer Angles

Prerequisites: The inscribed angle theorem

This activity serves as a review and a "real world" application of the inscribed angle theorem. Note that the word *locus* has been replaced by the word *location*. This has the advantage of not terrifying the students.

To discuss Questions C–E, you should hand out the Soccer Discussion sheet or make a transparency of it (or both).

The most difficult question by far is E. One way to think of it is to imagine two runners, W and C. As W runs down L_4, C stays at the center of the corresponding shooting arc. Since W's angle is half of C's, W has a better shot whenever C has a better shot. However, there comes a point where C must start moving back away from the goal, thereby reducing his shooting angle. This point corresponds to the time both runners are at the same horizontal line.

A challenging follow-up is to locate that point precisely by compass and straightedge construction. Another is to calculate its distance from the goal line as a function of the distance between the two runners' lines.

Answers

1–5. See student work.

6. The one corresponding to the 90° shooting angle

7. See student work.

8. 80°

9. The shooting angle for a person standing at the center of a circle is double the shooting angle for a person standing on the circle itself.

0. It is where the line of centers meets the 90° arc.

Discussion Answers

A. On the same circle, behind the goal

B. See the answer to Problem 9.

C. Where L_0 meets L_1

D. Where L_2 meets the goal line

E. On L_3, it would be at the point of tangency with the 40° circle. On L_4, it would be at the point of tangency with some shooting circle, with shooting angle between 20° and 30°. See teachers' notes above for an explanation.

2 Tangrams

Lab 2.1: Meet the Tangrams

Timing: This activity could take a whole period. If you don't want to give it that much time, do not duplicate this sheet, but address the basic ideas in Problems 4–6 and Questions A and B without it before going on to the next lab.)

Hand out the tangrams, one bag per student. To avoid mixing up the sets, do not give the same color to neighbors. Tell them that there are seven pieces per set and warn that, at the end of class, seven pieces should find their way back into the bag. At first, you should allow your students to play freely with the tangrams. Students may try to match the given figures or create their own. Houses, trees, and rocket ships are not uncommon. Students may record favorite designs by tracing the blocks. By tracing the outlines of their designs, students can create puzzles for their classmates to solve.

Some students may know how to make a square with all seven pieces, and others may want to try it. You should neither encourage nor discourage that activity. Emphasizing it too much early on can be demoralizing because it is quite difficult. You should discourage students from "giving away" the solution, which is shown for your reference.

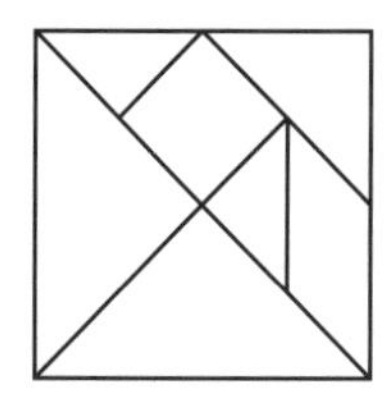

In Problem 4, make sure the students recognize that all five triangles are half-squares (isosceles right triangles). Introduce the word *parallelogram* and its spelling.

Understanding that the parallelogram is *not* flip-symmetric (see Problem 5) is sometimes the key to solving difficult tangram puzzles.

Question A hints at the notion of similarity. We will come back to this in Section 10, but if your students have already been exposed to this concept, you can ask them to make pairs of similar figures using tangrams. The figures below are examples.

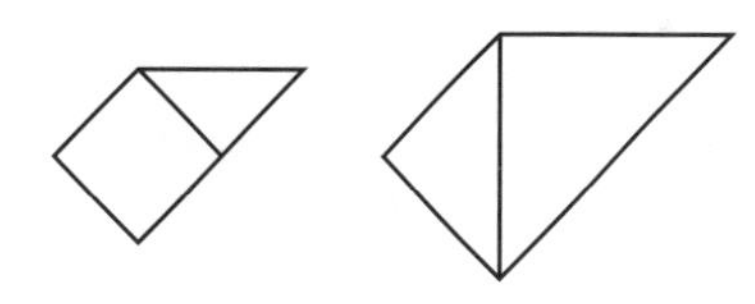

Answers

1. 7
2. See student work.
3. See student work.
4. Five right isosceles triangles, one square, and one parallelogram
5. All do except the parallelogram.
6. Triangles: 45°, 90°; square: 90°; parallelogram: 45°, 135°
7. Answers will vary.

Discussion Answers

A. They have the same shape but come in three different sizes. They are similar.

B. Two small triangles can be combined to make a square or a parallelogram.

C. Triangles: 360°, a whole circle; square: 90°, a quarter circle; parallelogram: 180°, a half circle

D. The figure is reorganized to produce a foot. Note that the figure with a foot has a slightly shorter body and smaller belly.

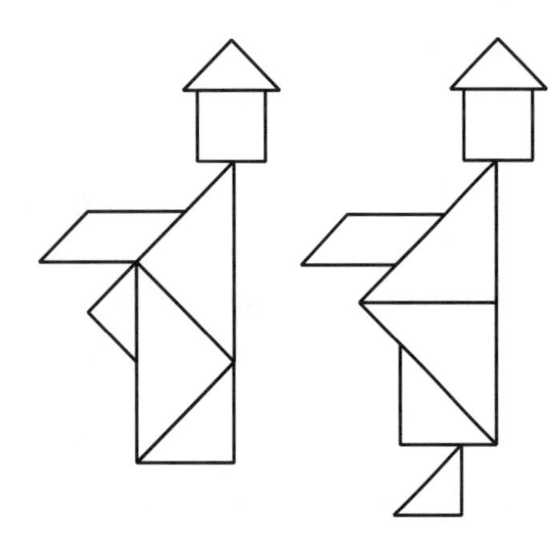

Lab 2.2: Tangram Measurements

This is a preview of an essential idea: the geometric interpretation of the square root. Understanding this idea makes it possible to solve the problem of finding all the tangram measurements. More important, it lays a foundation for a deeper understanding of square roots and their arithmetic than is possible from algebraic arguments alone. We will return to this with more detail and generality in Section 9.

Note that the Pythagorean theorem is not required for this lab.

In most classes, few students will be able to complete Problem 1 without the help of the questions in Problems 2–5. In fact, even after answering those questions, students may still need help. For example, the notation $2\sqrt{2}$ for the sides of the large triangles will not spontaneously occur to most students. Be prepared to introduce it and explain it.

Problem 7 can be assigned as an assessment of this lab.

All the discussion questions are difficult and deserve some time:

- Question A focuses on another important payoff of this lab: an understanding of the dimensions of the isosceles right triangle.
- Question B is a preview of similarity and scaling, which we will return to in Section 10.
- Question C gets at the idea—not at all obvious—that $2\sqrt{2} = \sqrt{8}$. The left side of that equation comes from measuring a leg of the large triangle with the legs of the small triangles. The right side comes from an argument about area similar to the one used in the lab to establish the length of a side of the square. You will find more on this subject in Section 9.

Of course, do not expect ideas such as these to be learned in a single lab. This activity is intended to complement, not replace, other approaches to these topics.

Answers

1. See answer to Problem 6 below.
2. **a.** 2 in.

   b. 2 sq in.

   c. 1 sq in.

   d. 2 sq in.
3. **a.** 81 sq units

   b. 3 units

   c. 25 sq units

   d. $\sqrt{5}$ units
4. **a.** s^2 sq units

   b. $\sqrt{A}$ units
5. $\sqrt{2}$ units
6. Small triangle sides: $\sqrt{2}$ in., 2 in. Area: 1 sq in.

   Medium triangle sides: 2, $2\sqrt{2}$ in. Area: 2 sq in.

   Square sides: $\sqrt{2}$ in. Area: 2 sq in.

   Parallelogram sides: $\sqrt{2}$ in., 2 in. Area: 2 sq in.

   Large triangle sides: $2\sqrt{2}$ in., 4 in. Area: 4 sq in.

7. Answers will vary.

Discussion Answers

A. $\sqrt{2}$

B. $\sqrt{2}, \sqrt{2}, 2$

C. 8 sq in.; $\sqrt{8} = 2\sqrt{2}$

Lab 2.3: Tangram Polygons

Timing: This activity could take one or two periods.

Part of the purpose is for students to develop their polygon vocabulary. The other part is to develop their visual sense, particularly to give them some experience with 45° angles and right isosceles triangles. Students who want to continue working on it beyond the allotted class time should be able to get extra credit or some sort of recognition for this. (You may set up a bulletin board display of the chart, in which you enter the names of students who solve a given puzzle.)

One way to organize this activity is explained below.

- Every time the students make a geometric figure, they quietly raise their hands. When you inspect their solutions, have them point to the appropriate space in their charts, indicating, for example, a three-piece triangle. If they are correct, initial that space. Some students may ask for more than one initial per space if they found different ways to solve a given problem. Some students may want you to check several spaces in succession before you leave to check other students' solutions.
- Help them label more lines in the chart beyond the initial three. If a student makes a three-piece rectangle, suggest starting a new row for rectangles. As more figures are found, you can update a list on the overhead or chalkboard so that other students can add to their sheets. Possible polygons include: rectangle, isosceles trapezoid, right trapezoid, pentagon, and hexagon.

This recording method is very hectic at first, as students start by finding easy solutions, but it gradually slows down as they start tackling more substantial challenges. Of course, you could simplify your life by having students check their own puzzles and by entering the names of the other polygons into the chart before duplicating it for the class. The reason I follow the above procedure is threefold:

- Students are not tempted to check off solutions they have not found.
- Getting your initial is more satisfying to many of them than their own check mark.
- Immature competition with classmates is curtailed because students' charts are individual, exhibiting different polygons or the same polygons in different order.

Many students set their own subgoals, such as finding all triangles or all figures that can be made with two, three, or seven pieces.

Note: The five-piece square turns out to be an important ingredient in solving some of the larger puzzles.

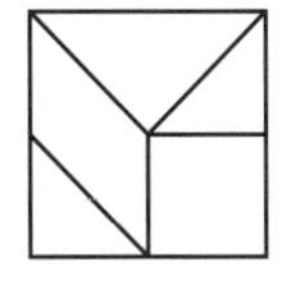

Answers

Answers will vary. See student tables and observe students as they work.

Discussion Answers

A. Answers will vary.

B. A single tangram piece has area 1, 2, or 4. All of them together have an area of 16. Therefore, a six-piece figure must have area 15, 14, or 12. It follows that a six-piece square is impossible, since its side would have to be the square root of one of these numbers. This is not possible given the measurements of the tangram pieces.

Lab 2.4: Symmetric Polygons

Prerequisites: Students need to have some familiarity with the concept of symmetry. If they do not, wait until you have done the opening lab of Section 5 before doing this one.

Problems 1 and 2 offer a good opportunity to think about what makes a figure symmetric.

The solutions to Problem 3 may well overlap with solutions found in Labs 2.1 (Meet the Tangrams) and 2.3 (Tangram Polygons). If you are short on time, skip this problem.

One approach to Question B is to work backward, starting from a symmetric figure and removing some pieces.

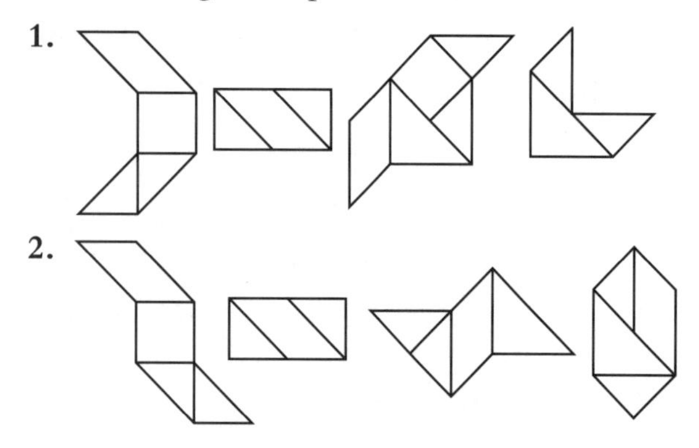

3. Answers will vary.

Discussion Answers

A. Answers will vary.

B. Answers will vary.

C. Yes. See the second and fourth figures in Problem 2, for example.

Lab 2.5: Convex Polygons

The formal definition of convex is:

A figure is convex if the segment joining any two points on it lies entirely inside or on the figure.

However, it is not necessary for students to understand this at this stage in their mathematical careers.

Problem 5 is quite open-ended. Its solutions overlap the ones found in Lab 2.3 (Tangram Polygons), but it is more difficult in some cases because convexity was not previously an issue.

Students probably have their own opinions on the aesthetics and difficulty of tangram puzzles. As I see it, there are three types of puzzles that are particularly pleasing:

- symmetric puzzles
- convex puzzles
- puzzles that use all seven pieces

The classic seven-piece-square puzzle satisfies all of these requirements. As for difficulty, it seems to also be related to convexity: If a figure has many parts "sticking out," it is easier to solve.

Extensions: Two classic seven-piece puzzle challenges are mentioned in Martin Gardner's book *Time Travel and Other Mathematical Bewilderment.*

- The more elegant of the two puzzles is to find all 13 seven-piece convex figures, which is mentioned in this lab as part of Problem 5.
- The other puzzle is to find as many seven-piece pentagons as possible.

Both challenges are very difficult and should not be required. I suggest using the first as an optional contest, displaying solutions on a public bulletin board as students turn them in, and perhaps offering extra credit. If this turns out to be a highly successful activity, you can follow it up with the other challenge. Do not tell students that there are 53 solutions to the pentagon challenge, as that would seem overwhelming. Instead, keep the search open as long as students are interested. The solutions can be found in Gardner's book.

Answers

1.

2–5. Answers will vary. See student work.

3 Polygons

Lab 3.1: Triangles from Sides

Prerequisites: This is the first construction activity in this book. It assumes that you have taught your students how to copy segments with a compass and straightedge. You may also use other construction tools, such as the Mira, patty paper (see next paragraph), or computer software. This lab, and most of the other construction labs, can be readily adapted to those tools or to some combination of them.

Patty paper is the paper used in some restaurants to separate hamburger patties. It is inexpensive and is available in restaurant supply stores and from Key Curriculum Press. It is transparent, can be written on, and leaves clear lines when folded. These features make it ideal for use in geometry class, as tracing and folding are intuitively much more understandable than the traditional compass and straightedge techniques. In my classes, I tend to use patty paper as a complement to a compass and straightedge, rather than as a replacement. (I learned about patty paper from Michael Serra. See his book *Patty Paper Geometry*.)

In content, the lab focuses on the Triangle Inequality. It is a useful preview of the concept of congruent triangles as well as a good introduction to the use of basic construction techniques.

Answers

1. See student work.
2. Answers will vary.
3. Sides a and b are too small, so they cannot reach each other if e is the third side.
4. **Possible:** *aaa, aab, abb, acc, add, ade, aee, bbb, bbc, bcc, bcd, bdd, bde, bee, ccc, ccd, cce, cdd, cde, cee, ddd, dde, dee, eee*

   Impossible: *aac, aad, aae, abc, abd, abe, acd, ace, bbd, bbe, bce*

Discussion Answers

A. No. Since 2 + 4 does not add up to more than 8.5, the two short sides would not reach each other.

B. No. Since 2 + 1.5 does not add up to more than 4, the two short sides would not reach each other.

C. It must be greater than 2 and less than 6.

D. The two short sides add up to more than the long side. Or: Any side is greater than the difference of the other two and less than their sum. (This is known as the Triangle Inequality.)

Lab 3.2: Triangles from Angles

Prerequisites: This lab assumes that you have taught your students how to copy angles with a compass and straightedge. As with the previous lab, you may also use other construction tools, such as the Mira, patty paper, or computer software. This lab, and most of the other construction labs, can be readily adapted to those tools or to some combination of them.

This lab reviews the sum of the angles of a triangle. It previews the concept of similar triangles and introduces basic construction techniques.

If it is appropriate for your class at this time, you may discuss the question of similar versus congruent triangles. The triangles constructed in this lab have a given shape (which is determined by the angles), but not a given size, since no side lengths are given. You could say that the constructions in the previous lab were based on SSS (three pairs of equal sides), while the ones in this lab are based on AAA, or actually AA (two—and therefore three—pairs of equal angles). SSS guarantees congruent triangles: All students will construct identical triangles. AA guarantees similar triangles: All students will construct triangles that have the same shape, but not necessarily the same size.

Answers

1. See student work.
2. Answers will vary.
3. No. The sides would not intersect.
4. **Possible:** Triangles can be constructed using the following pairs of angles: 1,1; 1,2; 1,3; 1,4; 1,5; 2,2; 2,3.

   Impossible: Triangles cannot be constructed using the following pairs of angles: 2,4; 2,5; 3,3; 3,4; 3,5; 4,4; 4,5; 5,5.

Discussion Answers

A. No. It's only possible if they add up to 180°.

B. They must add up to less than 180°.

C. **Method 1:** Construct a triangle that has those two angles. The third angle is the required size.

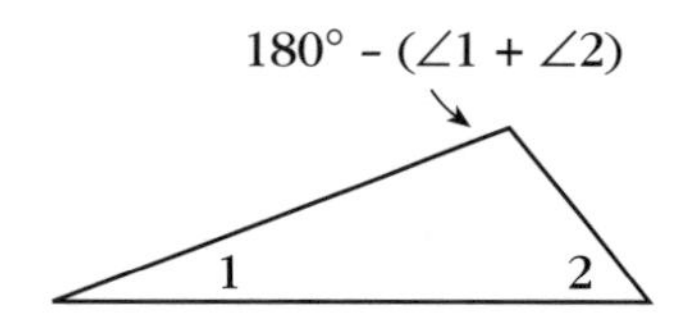

Method 2: Construct a copy of one of the angles so it is adjacent to the other. Extend either nonshared side to form a third angle, which, added to the first two, forms a straight line.

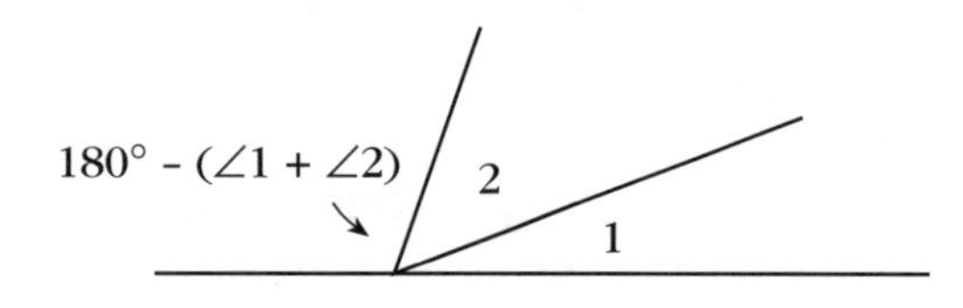

Lab 3.3: Walking Convex Polygons

The definition of convex polygon given here—a convex polygon is one where no angle is greater than 180°—is equivalent to this more formal one, given in the Notes to Lab 2.5 (Convex Polygons): A figure is convex if the segment joining any two points on it lies entirely inside or on the figure. You can convince yourself of this by drawing a few pictures. It is not necessary to discuss this with your students; they will be glad to accept the equivalence without proof. The alternate definition is provided here in case you skipped Lab 2.5 and you need a definition that is easy to grasp so you can get on with the work at hand. (See the introduction to Section 1 for some notes on angles greater than 180°.)

Some of the solutions to Problem 3 are quite difficult to find. It is not crucial for all students to find every one; you may ask them to work on this in groups. The main point of the exploration is to make sure students understand the definition of a convex polygon.

Solutions to this puzzle can be beautiful, and you may want to display them on a bulletin board. Students who enjoy this challenge can be given additional constraints: It's possible to solve the puzzle using only squares and triangles. Another challenge could be to use as many tan pattern blocks as possible, or to find alternative solutions for each n-gon.

The exterior angle was introduced for triangles in Lab 1.6 (The Exterior Angle Theorem), but that lab is not a prerequisite to this one.

Make sure students understand the example of walking the trapezoid before letting them work on other figures. You may have a student demonstrate the walk in a figure like the one on the sheet, while another records each step on the board or overhead projector.

If the trapezoid instructions start in the middle of a side, as suggested in Question A, all turn angles must be included in order to finish the walk. This provides another reason for including the final turn in the "normal" walks that start at a vertex: This way, all versions of the walk include the same angles.

Answers

1. No. All three angles combined add up to 180°, so one angle alone cannot be greater than 180°.
2. Yes. Figures will vary.

3. Answers will vary. Here are some possibilities.

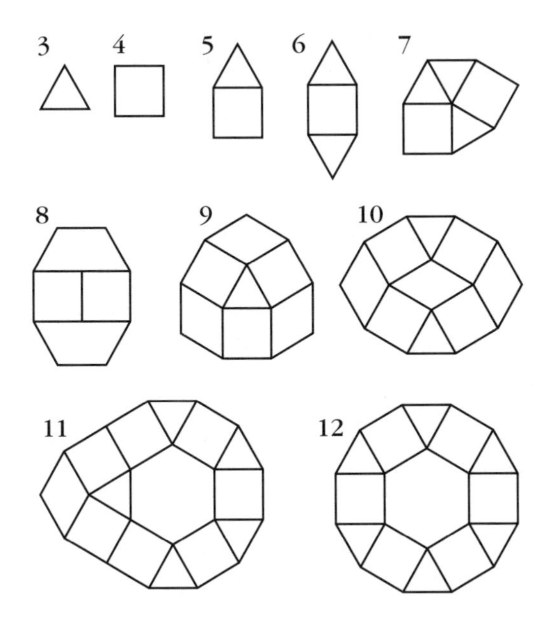

4. See figure below.

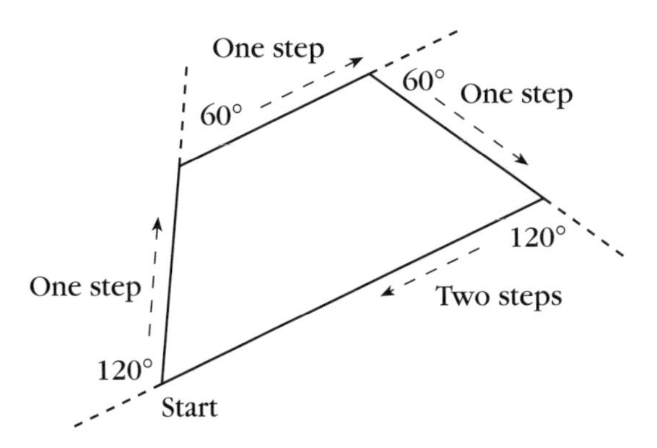

5. Answers will vary.
6. Answers will vary.
7. No written answer is required.
8. Answers will vary.

Discussion Answers

A. Answers will vary.

B. They are supplementary. They are equal if they are right angles.

C. In a clockwise walk, you turn right. In a counterclockwise walk, you turn left.

Lab 3.4: Regular Polygons and Stars

Prerequisites: Students should know how to find the measure of an inscribed angle given its intercepted arc. See Lab 1.7 (Angles and Triangles in a Circle) and Lab 1.8 (The Intercepted Arc).

To make sure students understand the definition of regular polygons, you may put a few pattern blocks on the overhead before handing out the geoboards, and precede the lab with a discussion of Questions A–C.

You may do this activity without actual geoboards or string by just using the circle geoboard paper on page 243. But it's fun to make at least a few of the stars on the geoboard using string. If you do use string, you will need to provide each student or pair of students with a string that is at least 5 or 6 feet long. For Question J, students can use the template's regular 10-gon to mark the vertices of the imaginary ten-peg circle geoboard. (In fact, students can use the template's regular polygons to investigate stars further.)

Questions E–J lead into a number of theoretic discussions about such things as common factors. Question I helps students generalize the calculation that they have to do while filling in the table. Questions J–L are wide-open opportunities to extend student research into this domain and could be the starting points of interesting independent or small-group projects.

Answers

1. Equilateral; square
2. **a.** A hexagon

   b. 120°
3. An eight-pointed star; 45°

4.

Every p-th peg	Star or polygon?	Number of sides	Angle measure
1	polygon	24	165°
2	polygon	12	150°
3	polygon	8	135°
4	polygon	6	120°
5	star	24	105°
6	polygon	4	90°
7	star	24	75°
8	polygon	3	60°
9	star	8	45°
10	star	12	30°
11	star	24	15°
12*	neither	2	0°
13	star	24	15°
14	star	12	30°
15	star	8	45°
16	polygon	3	60°
17	star	24	75°
18	polygon	4	90°
19	star	24	105°
20	polygon	6	120°
21	polygon	8	135°
22	polygon	12	150°
23	polygon	24	165°
24*	neither	0	180°

*Accept different reasonable answers for these rows and encourage students to defend their answers. See answer to Question G below.

Discussion Answers

A. Triangle, square, hexagon

B. The angles are not all equal.

C. Answers will vary.

D. Answers will vary.

E. You get a polygon only if p is a factor of 24 (excluding 12 and 24; see Question G).

F. Connecting every eleventh peg (or thirteenth peg) requires the most string because this produces a 24-sided star in which each side is nearly the diameter of the geoboard. Connecting every ninth peg (or fifteenth peg) requires the least string because this produces a star with only eight sides.

G. When $p = 12$ or $p = 24$, the figure would have two overlapping sides, or none, which would not be enough sides for either a polygon or a star. It is possible to find the angles by extending the patterns that are apparent in the other rows.

H. Every interval of p pegs counterclockwise is the same as an interval of $24 - p$ pegs clockwise.

I. $180° - 15p$

J. See student work.

K. Values of p that are factors of n will produce a polygon. Other values of p produce stars.

L. Answers will vary.

Lab 3.5: Walking Regular Polygons

Prerequisites: Students need to be familiar with the definition of interior versus exterior angles. These concepts were introduced in Lab 1.6 (The Exterior Angle Theorem) and Lab 3.3 (Walking Convex Polygons).

Actual students wrote the three samples at the beginning of the lab. You may try to work from samples written by your students. The first instruction is ambiguous and could be clarified either by writing "Turn 90° left and take a step. Repeat four times." or by using parentheses, as Maya did.

Students will probably not need to physically walk these shapes. If they did Lab 3.3 (Walking Convex Polygons), they should have little trouble doing the "walking" on paper. The essential idea, which Question B hints at, is that by the time you have walked all around the polygon, your total turning is 360°. All the other calculations follow easily from this fact.

If you use spreadsheets or graphing calculators, you can have students use them to complete the table; however, that requires figuring out the formulas. Another approach is to finish Problem 6 on paper, then extend the table by using technology.

Question C: The angle sum formula that flows from Problem 6 is $n(180° - 360°/n)$. This simplifies to $180°n - 360°$, which can be factored to the traditional $180°(n - 2)$.

Question D can be the starting point of an ambitious mathematical research project for a student, group, or class.

If you have access to the Logo computer language, your students may follow up this activity with the equivalent one on the computer. There, they can explore many variations of polygon walks. (See almost any of the books on Logo.) For further investigations, you could have students use Logo to write procedures to draw each of the pattern blocks. A more advanced project is to combine such procedures to create pattern block designs on the screen.

Answers

1. Jenny: instructions are ambiguous; Maya: 360°; Pat: 270°
2. (Take step, turn right 120°) · 3
3. 360°
4. **a.** The interior angles

   b. The exterior angles
5. **a.** Since the polygon is regular, all interior angles are equal. Therefore, all exterior angles are equal, and each is equal to the total turning divided by 7. So each exterior angle is 360°/7, or approximately 51.43°.

   b. Each interior angle is the supplement of the corresponding exterior angle. In this case, it is 180° − 51.43° = 128.57°.
6.

Number of sides	Each angle	Angles sum	Turn angle	Total turning
3	60°	180°	120°	360°
4	90°	360°	90°	360°
5	108°	540°	72°	360°
6	120°	720°	60°	360°
7	128.57°	900°	51.43°	360°
8	135°	1,080°	45°	360°
9	140°	1,260°	40°	360°
10	144°	1,440°	36°	360°
11	147.27°	1,620°	32.73°	360°
12	150°	1,800°	30°	360°
100	176.4°	17,640°	3.6°	360°
n	$180° - \frac{360°}{n}$	$n\left(180° - \frac{360°}{n}\right)$	$\frac{360°}{n}$	360°

Discussion Answers

A. **a.** Jenny's instructions are ambiguous. The other two will yield a square.

b. They all involve taking steps and turning 90°.

c. Maya's involve four turns; Pat's only three.

d. Answers will vary.

e. Answers will vary.

B. Students should agree that total turning is easiest to find, since it's always 360°. From there you can get the turn angle, the interior angle, and the angles sum, in that order. (See the formulas in the table above.)

C. Answers will vary.

D. Answers will vary.

Lab 3.6: Walking Nonconvex Polygons

To walk a nonconvex polygon requires at least one turn to be in a different direction from the others. For example, in the figure below, there are three right turns and one left turn if you go clockwise around the figure.

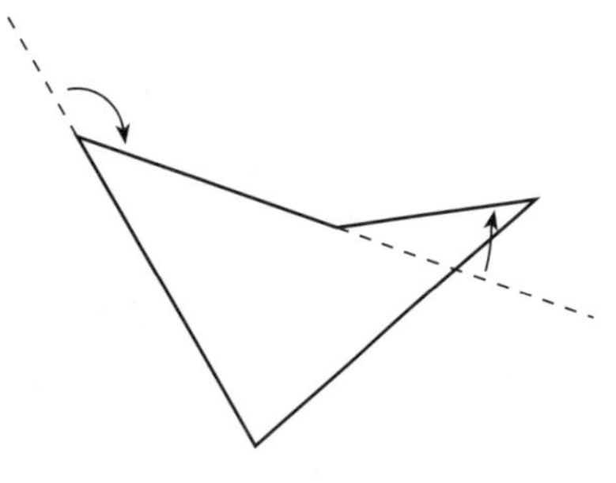

This leads to interesting questions about how to add the angles and still get 360°. You may ask the students to do Problems 1 and 2 and discuss Question A before handing out the sheet.

The idea of *heading,* introduced in Problem 2, gives a conceptual anchor to the idea of positive and negative turns. If you are complementing the walking lessons with work in the Logo computer language, you should be aware that in this lesson heading is defined the same way as in Logo: In Logo, there are no references to compass directions, but 0° is toward the top of the screen, 90° is toward the right, and so on.

Answers

1. **a.** Answers will vary.

   b. No written answer is required.

2. 360°

   a. Answers will vary.

   b. Answers will vary.

3. **a.** 180°

   b. 270°

   c. 225°

   d. 337.5°

4. **a.** 192°

   b. 303°

   c. 180° + h°. If the answer is greater than 360°, subtract 360°.

5. See answer to 4c.
6. Add 360° to get the corresponding positive heading.
7. Answers will vary.
8. Left turns subtract; right turns add.
9. If you consider right turns to be positive and left turns to be negative, the total turning should be 360° (assuming you are walking in an overall clockwise direction).

Discussion Answers

A. See answer to Problem 9 above.

B. Turning 360° brings your heading back to what it was when you started. If you turn 350°, you have turned 10° less than all the way around, which is the same as turning −10°.

C. Right −90° = left 90° = right 270°

D. Clockwise

Lab 3.7: Diagonals

In Problem 3, encourage students to come up with a systematic method for counting. One way, which you may suggest if they are getting frustrated, is to count the number of diagonals out of one vertex, then count the number of *new* diagonals out of the next vertex (one less), and so on.

One way to get at the formula, which you may suggest as a complement to your students' approaches, is to connect each vertex to all other vertices not adjacent to it. In an n-gon, there are $n - 3$ vertices that are nonadjacent to any given vertex; so there are $n - 3$ diagonals per vertex, for a total of $n(n - 3)$. However, this method counts every diagonal twice, once at each end. Therefore, the actual number is $n(n - 3)/2$. This reasoning is relevant to Problems 3–5.

To help students understand the comment following Problem 4, you should probably show examples of external diagonals in nonconvex polygons on the overhead or chalkboard.

To make an algebra connection, you may have the students make a graph of the data in Problem 3. When they find the formula, they may use a graphing calculator to see that the graph actually does go through the data points.

The discussion in Problems 5–7 continues a thread we started in Lab 1.4 (Angles of Pattern Block Polygons) and continued in Lab 3.5 (Walking Regular Polygons). Here we present a traditional approach to finding the sum, but it requires the polygon to be convex.

Answers

1. 0
2. 2
3.

Sides	Diagonals
3	0
4	2
5	5
6	9

Sides	Diagonals
7	14
8	20
9	27
100	4850

4. $\frac{n(n-3)}{2}$
5. $n - 2$
6. $(n - 2)180°$
7. External diagonals do not divide the polygon into triangles.

Discussion Answers

A. Successive differences in the numbers of diagonals are 2, 3, 4, and so on.

B. A polygon is convex if all of its diagonals are on the inside of it.

Lab 3.8: Sum of the Angles in a Polygon

This is another lesson on the same topic. The point of these multiple approaches is not so much to stress the formula—which is not in itself earth-shatteringly important—but it is interesting to see that there are many avenues to get to it. Since this lab requires no materials and travels familiar territory, it could be done as homework.

Answers

1. **a.** Answers will vary.
   b. Answers will vary.
   c. 720°
   d. 360°
   e. 360°
2. **a.** 540°, 720°, 1080°, 1800°
   b. $180°n - 360°$
3. The procedure fails because no interior point can be connected to all the vertices.
4. Answers will vary.
5. The sum should be 900°.

Discussion Answer

A. This method uses two more triangles but yields the same formula.

Lab 3.9: Triangulating Polygons

If your students are familiar with Euler's formula for networks ($E = V + F - 1$, where V, F, and E are the number of vertices, "faces" [that is, undivided polygons], and edges, respectively), you may investigate the relationship between that formula and the one in this exploration.

Lead a discussion of Question D. One possible answer is that the sum of the angles in a polygon can be found by triangulating it. In fact, that is the method we used in the previous two activities, but only now do we have a method general enough to work for any polygon, convex or not.

Duane W. DeTemple and Dean A. Walker describe an interesting set of activities that make use of polygon triangulation in "Some Colorful Mathematics" (*The Mathematics Teacher,* April 1996, p. 307).

Answers

The following answers correspond to the list of research questions.

- One triangle is added each time you add a side.
- Adding an inside vertex adds two triangles.

- Adding a side vertex adds one triangle.
- The relation can be expressed as $t = n - 2 + 2i + s$.

Discussion Answers

A. No.

B. A minimal triangulation might be defined as a triangulation that requires no additional vertices besides the vertices of the polygon. It can be sketched by drawing as many inside diagonals as possible, with the restriction that they should not intersect each other.

C. Inside vertices add more triangles than do side vertices. Four original vertices, two side vertices, and two inside vertices will yield eight triangles, whereas the example with six original vertices, one inside vertex, and one side vertex yields only seven triangles. The additional triangle is a result of the additional inside vertex. If there are eight vertices, the way to get the least possible number of triangles is to make all eight vertices the original vertices of an octagon. Then there are only six triangles. The way to get the greatest possible number of triangles is to start with a triangle for the original shape and add five inside vertices. That gives a total of 11 triangles.

D. Triangulation provides a method for calculating the sum of the angles in a polygon.

4 Polyominoes

Lab 4.1: Finding the Polyominoes

This lab is a prerequisite to all further work with polyominoes and pentominoes, since this is where we define our terms and discover and name the figures.

Throughout this section, use 1-Centimeter Grid Paper (see page 241).

The main difficulty is that students tend to discover the same pentominoes more than once. The best way to eliminate duplications is to make the polyominoes with the interlocking cubes. Then students can pick up, turn, and flip newly found pentominoes to check whether they are identical to shapes they have already found. To avoid finding 3-D shapes, I include the rule that when the figure is laid flat all of its cubes should touch the table. The 3-D version of this activity is polycubes (see below).

You will have to decide whether to reveal the total number of tetrominoes (five) and pentominoes (12) or to let your students discover them on their own. I have done both, depending on how much time I had for the activity and whether any of the students seemed ready for the somewhat more demanding task of developing a convincing argument that all the figures have been found.

Later in this section you will find other searches in the same style, though generally more difficult: hexominoes, polyrectangles, polycubes, and polyarcs.

Here are some suggestions for the alternative classification schemes referred to in Question B (you can find more later in the section):

- by perimeter (for these areas, the result would be almost the same as the classification by area, but not quite; see Section 8)
- by number of sides (interestingly, always an even number—why?)
- convex versus nonconvex (this reveals that the only convex polyominoes are rectangles)

After this lesson, you can hand out the Polyomino Names Reference Sheet that follows the lab. Encourage students not to lose their copies, as you may need to discuss the polyominoes with specific references to various names. (It is not, however, important to memorize the names.)

Answers

1. See Polyomino Names Reference Sheet.
2. See Polyomino Names Reference Sheet.

Discussion Answers

A. Answers will vary.

B. See notes above.

C. F, L, N, T, W, X, Y, Z

Lab 4.2: Polyominoes and Symmetry

This lab is a preview of some of the ideas about symmetry discussed in Section 5, approached from the point of view of polyominoes. The use of a grid here makes it somewhat easier to introduce these ideas, but this lesson is not a prerequisite for Section 5.

If students find Problem 2 daunting (after all, there are 20 categories!), suggest that they go through the list of polyominoes (see Polyomino Names Reference Sheet) and find the appropriate category for each one. For most students, this would be easier than searching for all the polyominoes that fit in each category.

We come back to the question posed here in Lab 4.10 (Polyrectangles), Question C.

Answers

1. Monomino: 1
   Domino: 2
   Bent tromino: 4
   Tetrominoes: square, 1; **l**, 8; i, 2; n, 4; t, 4
   Pentominoes: F, 8; L, 8; I, 2; P, 8; N, 8; T, 4; U, 4; V, 4; W, 4; X, 1; Y, 8; Z, 4
2. **a.** No mirrors, no turns: tetromino **l**; pentominoes F, L, P, N, Y

   b. One mirror, no turns: bent tromino; tetromino t; pentominoes T, U, V, W

   c. Two mirrors, no turns: none

   d. Three mirrors, no turns: none

   e. Four mirrors, no turns: none

   f. No mirrors, two-fold turn (180°): tetromino n; pentomino Z

   g. One mirror, two-fold turn: none

   h. Two mirrors, two-fold turn: domino; straight tromino; tetromino i; pentomino I

   i. Three mirrors, two-fold turn: none

   j. Four mirrors, two-fold turn: none

   k. No mirrors, three-fold turn (120°): none

   l. One mirror, three-fold turn: none

   m. Two mirrors, three-fold turn: none

   n. Three mirrors, three-fold turn: none

   o. Four mirrors, three-fold turn: none

   p. No mirrors, four-fold turn (90°): none

   q. One mirror, four-fold turn: none

   r. Two mirrors, four-fold turn: none

   s. Three mirrors, four-fold turn: none

   t. Four mirrors, four-fold turn: monomino; tetromino square; pentomino X

Discussion Answers

A. Three-fold turns are impossible because all angles are 90°. Other impossible combinations are more difficult to explain; for example, if there are two mirrors, there has to be a turn.

B. The more symmetric the shape, the fewer positions there are on graph paper.

C. You would need to have backward versions of the tetromino **l** and the following pentominoes: F, L, P, N, Y.

Lab 4.3: Polyomino Puzzles

For Problem 2, you may choose to reveal that there are 25 such rectangles. For Problem 4, there are six such staircases.

Note that the area of a rectangle of the type requested in Problem 2 cannot be a prime number and that the process of finding the rectangles is equivalent to finding the factors of the nonprime numbers from 4 to 28.

Answers

1. See Polyomino Names Reference Sheet.

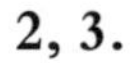

2, 3.

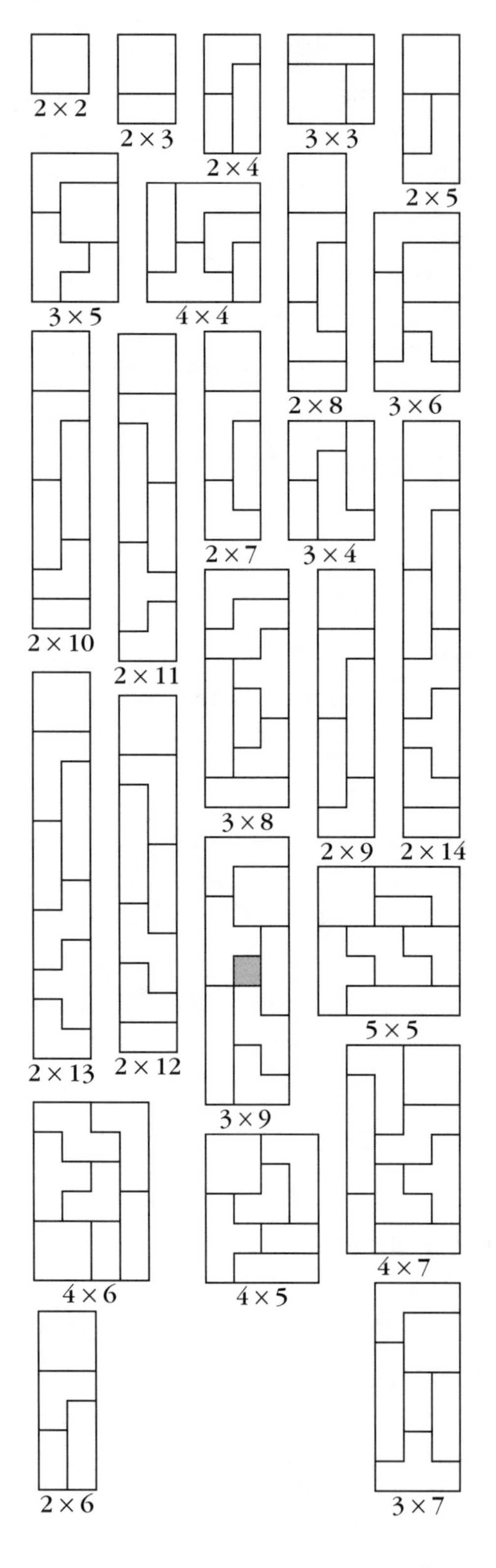

4.

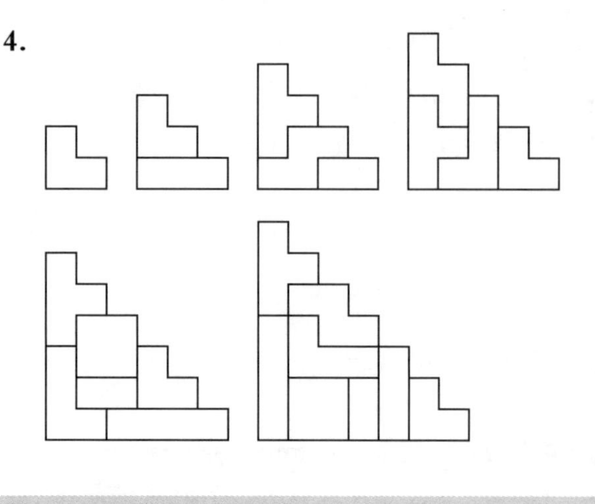

Discussion Answers

A. Answers will vary.

B. The 3×9 rectangle is impossible because its area is 1 less than the combined area of the allowed pieces. However, the piece with least area is the domino, and omitting it allows a maximum area of 26.

C. Two such staircases can be combined to make an $x \times (x + 1)$ rectangle. So the area of one must be $x(x + 1)/2$.

Lab 4.4: Family Trees

Family trees and envelopes (see Lab 4.5) offer ways to make the search for hexominoes more systematic. In addition, they offer yet two more methods for classifying polyominoes.

If your students have studied probability, you could assign the following research project.

Imagine that a machine creates pentominoes with this algorithm: Start with a monomino and add squares randomly, with each possible location having an equal chance of being used. What is the probability of getting each pentomino? Which pentomino is the most likely? Which is the least likely?

It is interesting to note that the P pentomino is the most likely, which is perhaps related to the fact that it is most often the one you need in order to complete pentomino puzzles.

Answers

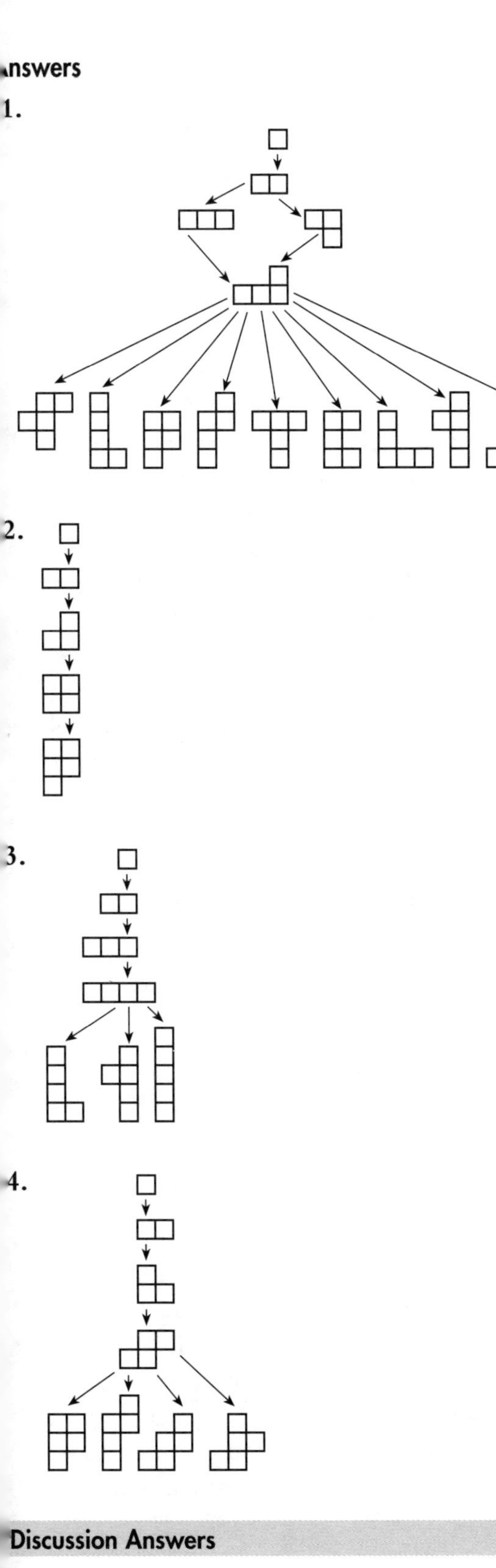

Discussion Answers

A. The P has four parents.

B. L and Y

C. F, N, P

D. Only the W

E. I and W

F. The L has 11 children.

G. The X has only two children.

Lab 4.5: Envelopes

If students are floundering on Problem 4, you may suggest that they organize their search with the help of the hexomino envelopes they found in Problem 3. Another useful approach is to use some sort of systematic process based on the family tree idea: Look for all of the children of each of the 12 pentominoes.

The advice given in Lab 4.1 (Finding the Polyominoes) still holds here.

Have students make the hexominoes with the interlocking cubes so that they can pick up, turn, and flip newly found hexominoes to check whether they are identical to shapes they have already found. Remind students that when the figure is laid flat all of its cubes should touch the table. The 3-D version of this activity is Lab 4.8 (Polycubes).

You will have to decide whether to reveal the total number of hexominoes (35) or to let your students discover it on their own. I have done both, depending on how much time I had for the activity and whether any of the students seem ready for the somewhat more demanding task of developing a convincing argument that all the figures have been found.

Because the task is so daunting, it is a good idea to work in cooperative groups. I have sometimes split the class into two groups (the boys against the girls!) and made this a race, but you may not have the tolerance for the level of noise and organized mayhem this may generate. If the groups are large, or even with normal-size groups of four, this is one exercise where it is important to have a student be responsible for record keeping. Among other things, this student must keep an eye out for duplicates!

It is not crucial that students find all 35 hexominoes, but they are likely to want to know if they were successful in their search. Use their curiosity as a springboard for the next lab, which should help them spot duplicates as well as find any missing hexominoes. In fact, if you do not want to spend too much time on this project, you may hand out the next lab early on.

Since there are 105 heptominoes, that search is probably best left to a computer.

Lab 4.7 (Minimum Covers) addresses a related question.

Answers

1. 1×4: i
   2×2: square
   3×2: n, l, t
2. 1×5: I
   2×3: P, U
   2×4: L, N, Y
   3×3: F, T, V, W, X, Z
3. | | | | |
   |---|---|---|---|
   | ~~2 × 2~~ | ~~1 × 5~~ | (2 × 3) | (2 × 5) |
   | (3 × 3) | ~~3 × 5~~ | (2 × 4) | (1 × 6) |
   | (3 × 4) | ~~2 × 6~~ | ~~4 × 4~~ | ~~3 × 6~~ |
4. √ = Can be folded into a cube

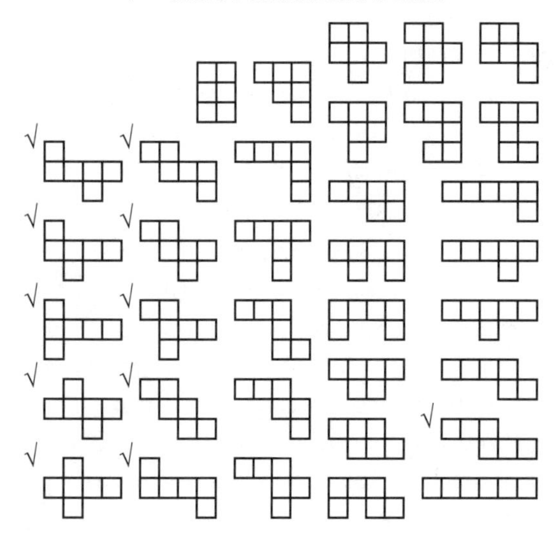

Discussion Answers

A. In most polyominoes, the number of vertical units on the perimeter equals twice the height of the envelope, and the number of horizontal units on the perimeter equals twice the width of the envelope. So the perimeter of the polyomino is equal to the perimeter of its envelope. For example, the n tetromino, oriented as shown below left, has 6 vertical units and 4 horizontal units along its perimeter. Its perimeter is 10, and its envelope is a 3×2 rectangle with perimeter 10. Polyominoes with "bays" in them can have perimeters greater than their envelopes. For example, the U pentomino has perimeter 12, but its envelope has perimeter 10. The perimeter around the inside of the "bay," 3 units, corresponds to just 1 unit on the perimeter of the envelope. It's not possible for a polyomino to have perimeter less than its envelope.

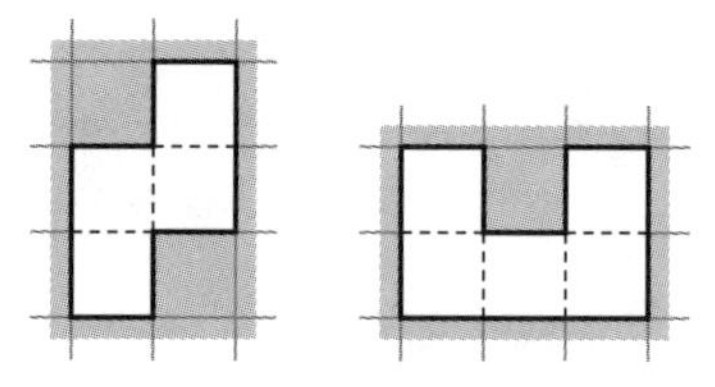

B. See check marks in the answer to Problem 4.

Lab 4.6: Classifying the Hexominoes

This lab is a sequel to Problem 4 in Lab 4.5. Students were asked to come up with envelopes and find hexominoes on their own, without knowing how many there are. This lab gives students the 35 hexomino envelopes, allowing them to check how they did in their search and to organize their findings. You can use this lab to speed up the investigation in Lab 4.5.

Answers

See answer to Problem 4, Lab 4.5.

.ab 4.7: Minimum Covers

This lab is a minimization problem, which hould help students develop their visual sense. t is not very difficult once the question is clear.

\nswers

1–3.

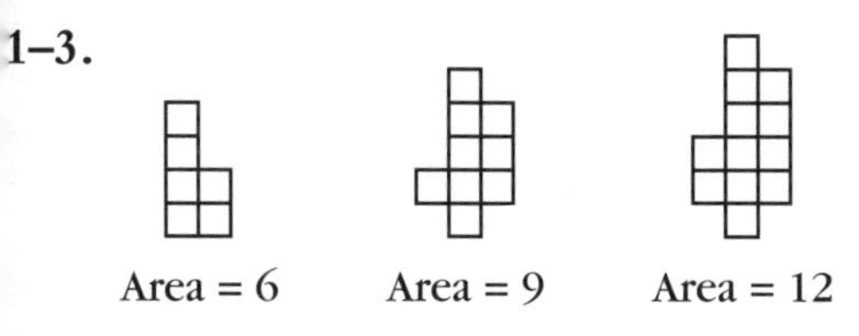

Discussion Answers

\. Answers will vary.

3. Other shapes are possible, though all minimum covers will have the same area. For example, this shape is also a minimum cover for tetrominoes.

.ab 4.8: Polycubes

'or many students, it is difficult to realize that here are 3-D figures that are mirror images of ach other yet cannot be superimposed onto each ther. This is the case whenever the figures are ot themselves mirror symmetric. This is unlike he situation in two dimensions, where (by way f a trip to the third dimension) a figure can be ipped and superimposed upon its mirror image.

ust as the canonical polyomino puzzle is o make a rectangle using a given set of olyominoes, the canonical polycube puzzle is to nake a box using a given set of polycubes. In 'roblem 5, one necessary condition for a box to vork as a pentacube puzzle is that its volume— nd therefore at least one of its dimensions— nust be a multiple of 5. Of course, that does not uarantee the puzzle can be solved (for example, 1 × 2 × 5 box cannot be made), but a olution can probably be found in almost every ase, if we are to generalize from the case of entominoes. This would be a good exploration or a student project.

To answer Questions C and D, students need to have done Lab 4.4 (Family Trees) and Lab 4.5 (Envelopes).

Look for other generalizations of polyominoes in the following two labs.

Answers

1.

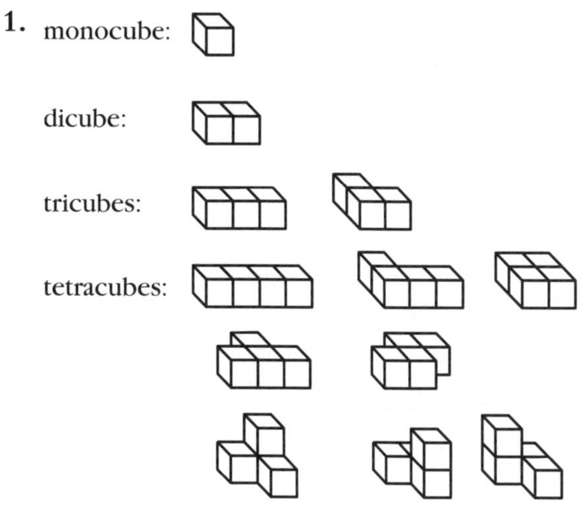

2, 3. No answer is required.

4. There are 32 possible pentacubes, including the 12 pentomino shapes.

5. Answers will vary. See Notes above.

Discussion Answers

A. Answers will vary.

B. By symmetry, by volume, by surface area, by convexity. Other answers are possible.

C. The smallest rectangular box that encloses the given polyomino

D. Answers will vary.

Lab 4.9: Polytans

Problem 2 is challenging, and you should encourage group collaboration to help those students who are getting stuck. You need to decide whether to tell the class how many distinct 4-tan shapes there are (14) or whether instead to have them discuss Question B.

Plastic 4-tan pieces are available commercially as SuperTangrams™, as are some related puzzle books I authored: *SuperTangram Activities* (2 vols., Creative Publications, 1986).

Answers

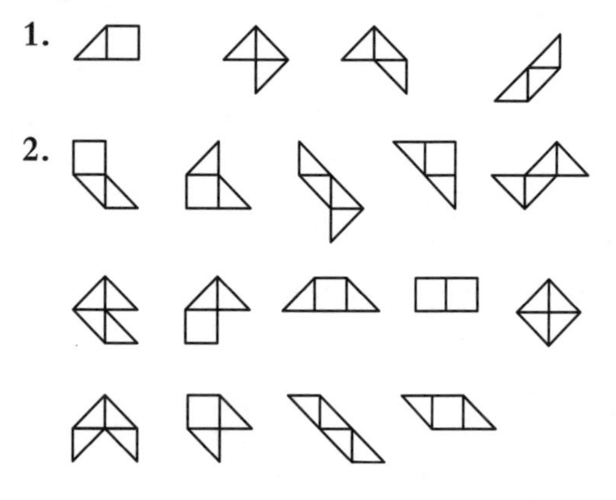

Discussion Answers

A. Answers will vary.

B. Answers will vary.

Lab 4.10: Polyrectangles

This is another generalization of the concept of polyomino, with a basic unit that is no longer a square. Another possible extension is the polyiamond (diamond, triamond, and so on), based on an equilateral triangle unit. Polyiamonds have been discussed by Martin Gardner (in his *Sixth Book of Mathematical Games from Scientific American,* chap. 18 and elsewhere).

5.

Discussion Answers

A. Polyominoes with a diagonal line of symmetry, such as the tetromino square or the pentominoes V, W, and X, yield only one polyrectangle.

B. Each polyrectangle yields one or two polystamps, depending on whether it is mirror symmetric or not. Each polystamp yields one or two one-positional polystamps.

C. False. The less symmetrical shapes yield more versions.

5 Symmetry

Lab 5.1: Introduction to Symmetry

This is a "getting ready" activity. Understanding the meaning of line and rotation symmetry is a prerequisite to all the labs in this section.

As a way to introduce these ideas, you may lead a discussion of the activity before handing out the sheet. One way to start such a discussion is to write some student-suggested letters on the chalkboard in two categories, "special" and "not special" (corresponding to "line symmetric" and "not line symmetric"), and see whether students can guess which other letters belong in each category. Shapes other than letters can also be used, of course, and the activity can be generalized to rotation symmetry, or made more specific by requiring criteria such as a horizontal line of symmetry or more than one line of symmetry.

In Question C, students should have no trouble classifying the figures according to their rotation symmetries, but it can be difficult to explain what is meant by "n-fold" rotation symmetry. One possible answer is that a figure has n-fold rotation symmetry if it looks exactly the same through n different turns, up to 360°. Another answer is that a figure with n-fold rotation symmetry looks unchanged when rotated by an angle of $360°/n$. Point out that any figure looks the same when it's rotated by 360°. If that's the only rotation that doesn't change the figure, then we say the figure does not have rotation symmetry.

Make sure your students understand that the word *fold,* in this context, is not related to folding. Students may also be misled into thinking *fold* refers to reflection symmetry only, especially since this lab mentions folding in Question B.

Note that the letters that have rotation symmetry all have half-turn symmetry. You might discuss whether X and O have rotation symmetries that other letters do not.

Bulletin board displays can be created based on the ideas in this activity, especially exhibits of discoveries made by students in working Problem 8. You may have an ongoing search, with new discoveries being added. Some students will probably enjoy looking for the longest possible answer in each. Many more examples of letter symmetries like the one in Problem 7 can be found in Scott Kim's book *Inversions* (Key Curriculum Press), the definitive book on the subject.

For Question F, if you don't have a dictionary in your classroom, you may ask your students to look up *symmetry* as part of their homework. *Webster's College Dictionary* (Random House) has this first definition: "the correspondence in size, form, and arrangement of parts on opposite sides of a plane, line, or point; regularity of form or arrangement in terms of like, reciprocal, or corresponding parts."

Answers

1. **a.** Answers will vary.

   b.

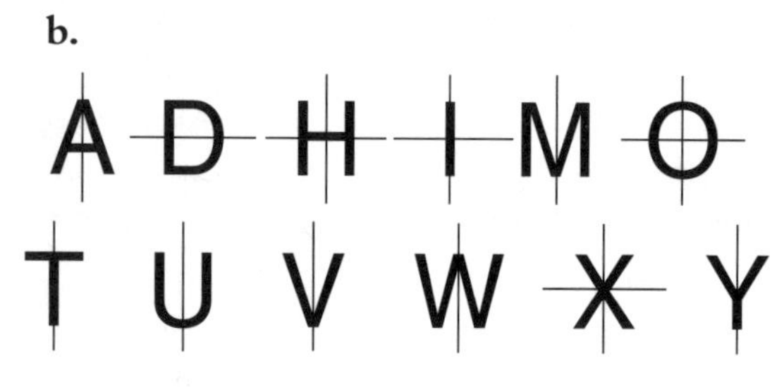

2. Line symmetric:

   B, C, E, K, Q

   Not line symmetric:

   B, C, E, K, Q

3. Line symmetric:

   c, i, l, o, v, w, x

   Not line symmetric:

   a, b, d, e, f, g, h, j, k, m, n, p, q, r, s, t, u, y, z

4. Answers will vary.

5. Rotation symmetry:

   H, I, N, O, S, X, Z

   No rotation symmetry:

   A, B, C, D, E, F, G, J, K, L, M, P, Q, R, T, U, V, W, Y

6. Rotation symmetry:

   l, o, s, x, z

   No rotation symmetry:

   a, b, c, d, e, f, g, h, i, j, k, m, n, p, q, r, t, u, v, w, y

7. The word HORIZON is written so that it has rotation symmetry. The H becomes an N when rotated 180° and the R becomes a Z.

8. Answers will vary.

Discussion Answers

A. If a mirror is placed along the line of symmetry, the reflection of one side in the mirror matches the other side. *Bilateral* mean two-sided, and line symmetric figures have two sides that are reflections of each other. Line symmetric figures, if flipped over, fit on their original outlines.

B. The two halves will cover each other exactly

C.

D. two-fold: infinity symbol, diamond; three-fold: recycling symbol; four-fold: pinwheel; five-fold: star; six-fold: asterisk. See Notes for explanations.

E. 180° is half of 360°, which would be a full turn. Figures with this kind of symmetry are symmetrical through their center point; that is, every point on the figure has a corresponding point the same distance from the center on the opposite side.

F. Answers will vary.

G. Among the capital letters: H, I, O, X. They all have more than one line of symmetry.

H. Answers will vary.

Lab 5.2: Triangle and Quadrilateral Symmetry

Problem 1 takes time, but it serves to consolidate students' grasp of the definitions. You may shorten it by asking for just one example of each shape, especially if your students are already familiar with the definitions.

You may lead a discussion of Question B on the overhead projector, entering appropriate nformation in the relevant cells of the table for Problem 4.

Although the list of quadrilaterals may appear to be ordered in some sort of mathematical scheme, actually it is in reverse alphabetical order.

Answers

1. Answers will vary.
2. Answers will vary.
3. Answers will vary.
4. See table below.

Discussion Answers

A. Yes. A figure with *n* lines of symmetry has *n*-fold rotation symmetry.

B. There are figures with three- and four-fold rotation symmetry, but no lines of symmetry; however, they are not triangles or quadrilaterals. Figures with exactly one line of symmetry cannot have rotation symmetry. Figures with exactly two, three, and four lines of symmetry must have two-, three-, and four-fold rotation symmetry, respectively.

C. Answers will vary. The scalene triangles have no symmetries. The isosceles figures have exactly one line of symmetry. If you join the midpoints of a rhombus, you get a rectangle, and vice versa.

D. The two lines of symmetry of the rhombus or rectangle are among the four lines of symmetry of the square. A square is a rhombus, but a rhombus is not necessarily a square. The line of symmetry of an isosceles triangle is among the three lines of symmetry of the equilateral triangle. An equilateral triangle is isosceles, but an isosceles triangle is not necessarily equilateral.

Lab 5.3: One Mirror

Prerequisites: Students should be familiar with the names of the triangles and quadrilaterals listed in the chart. In this book, those names were introduced in Lab 1.5 (Angles in a Triangle) and Lab 5.2 (Triangle and Quadrilateral Symmetry). In this activity, always use the most generic version of a figure: *isosceles* means a triangle with exactly two equal sides, *rectangle* means a rectangle that is not a square, and so on.

If you have never used mirrors with your class, you should consider giving students a little time to just play freely with them at the start of the lesson. Then, you may discuss the following examples as an introduction.

	Rotation Symmetry			
Line Symmetry	**None**	**Two-fold**	**Three-fold**	**Four-fold**
No lines	AS, RS, OS, HE, GT, GQ	PA		
One line	AI, RI, OI, KI, IT			
Two lines		RH, RE		
Three lines			EQ	
Four lines				SQ

By placing the mirror on an equilateral triangle, students can make an equilateral triangle, a kite, or a rhombus, as shown in this figure.

Equilateral ⟶ Equilateral, Kite, Rhombus

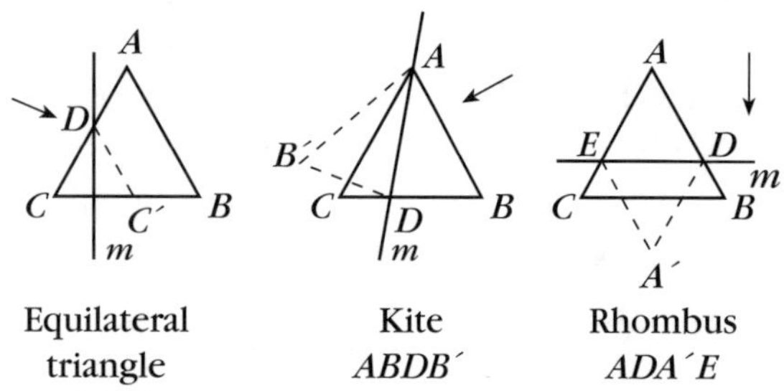

Equilateral triangle *DCC′* Kite *ABDB′* Rhombus *ADA′E*

(*m* is the mirror—the arrow indicates the side you look from)

They cannot, however, make an acute isosceles triangle, since it is impossible to get the needed non-60° angles.

Students may work in pairs or groups to fill out the chart.

Make sure they use all the possible triangles on the template in their search for solutions. Something may be impossible with one isosceles triangle yet work with another one. (An obtuse isosceles triangle can be obtained from an acute isosceles triangle only if the vertex angle is less than 45°.) This may yield some heated discussions!

Answers

See chart below.

Figures made	By using the mirror on:							
Triangles	**EQ**	**AI**	**RI**	**OI**	**AS**	**RS**	**HE**	**OS**
Equilateral (EQ)	✓	×	×	×	×	×	✓	×
Acute isosceles (AI)	×	✓	×	×	✓	✓	×	✓
Right isosceles (RI)	×	×	✓	×	×	×	×	×
Obtuse isosceles (OI)	×	✓	×	✓	✓	✓	✓	✓
Acute scalene (AS)	×	×	×	×	×	×	×	×
Right scalene (RS)	×	×	×	×	×	×	×	×
Half-equilateral (HE)	×	×	×	×	×	×	×	×
Obtuse scalene (OS)	×	×	×	×	×	×	×	×
Quadrilaterals	**EQ**	**AI**	**RI**	**OI**	**AS**	**RS**	**HE**	**OS**
General	×	×	×	×	×	×	×	×
Kite	✓	✓	✓	✓	✓	✓	✓	✓
General trapezoid	×	×	×	×	×	×	×	×
Isosceles trapezoid	×	×	×	×	×	×	×	×
Parallelogram	×	×	×	×	×	×	×	×
Rhombus	✓	✓	✓	✓	✓	✓	✓	✓
Rectangle	×	×	×	×	×	×	×	×
Square	×	×	✓	×	×	×	×	×

Discussion Answers

A. Pentagons and hexagons

B. It is impossible to make any figure that is not line symmetric. This follows from the fact that any figure made with a mirror is automatically line symmetric.

C. They are line symmetric.

D. It is impossible to make the isosceles trapezoid and the rectangle, because the half-figure that would be needed to reflect in the mirror has two opposite parallel sides and therefore cannot be part of a triangle.

E. The kite and rhombus can be made from any triangle, because putting a mirror on any angle will yield a kite unless the mirror is perpendicular to one side (see Question G) or perpendicular to the angle bisector. In the latter case, it makes a rhombus.

F. Find the needed half-figure, and place the mirror along it.

G. Place the mirror perpendicular to a side so that side yields only one side in the final figure.

H. An acute angle in an obtuse isosceles triangle is less than 45°. If it is reflected in the mirror, its reflection will also be less than 45°. Together, they will add up to less than 90°, leaving more than 90° for the vertex angle. On the other hand, if the vertex angle in an acute isosceles triangle is less than 45°, then by the same reasoning, reflecting that angle in the mirror will create an obtuse angle for the new vertex angle, thereby making an obtuse isosceles triangle.

Lab 5.4: Two Mirrors

For this lab, students will need to attach two mirrors to each other so they form a hinge. Give the following instructions, demonstrating as you go along: Start by placing the two mirrors face to face, then tape them together along a short side. If you do this carefully, you should be able to swing them open to any angle. Unless the angle is close to 0° or 180°, the hinged mirror pair should be able to stand up as shown here.

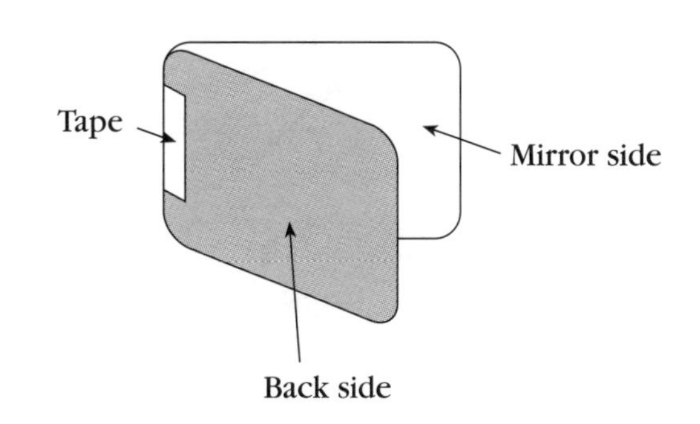

If students don't start with the mirrors face to face, the resulting contraption will not work as well. You should rule out the use of tape on the front of the mirror, as it would mar the surface.

If your students prefer to have their own mirrors to work with, rather than sharing the hinged pairs between two students, you can separate the mirrors after this lab. Otherwise, you may keep the mirrors taped together; the hinged pairs will be useful (though not necessary) in Lab 5.7 (Two Intersecting Lines of Symmetry). You can also just use the pair as a free-standing single mirror by using a very large angle (say 300°).

This lab is very rich, and potentially complex. The student page concentrates on some basic results, which Questions A–G expand to a variety of related questions. Give yourself time to explore all this on your own first, then decide how deeply you want to get into it with your class. Much of the complication stems from the fact that mirrors and theoretical lines of symmetry are not exactly the same thing: An actual mirror has a beginning and an end and reflects only in one direction, while a theoretical line of symmetry is infinite and works in both directions.

Students may use Isometric Dot Paper (see page 246) as a shortcut for getting 120°, 60°, and 30° angles.

In addition to record keeping, the template may be used to draw figures that can be reflected in the mirror pair. Drawn figures have the advantage that the mirrors can be placed right over the figures.

Answers

1.

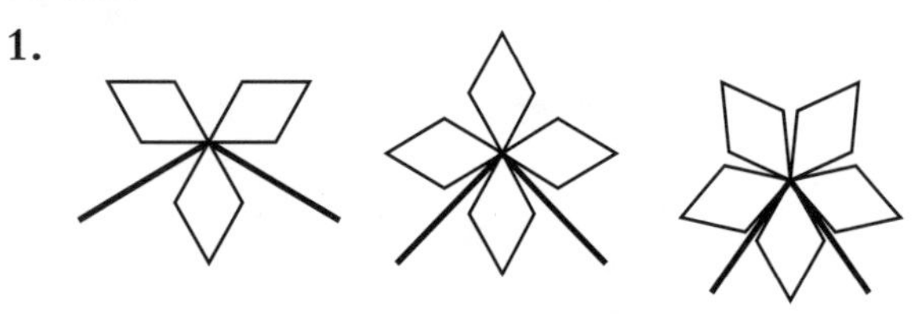

2. 120°, 90°, 72°. These are the results of dividing 360° by 3, 4, and 5.

3.

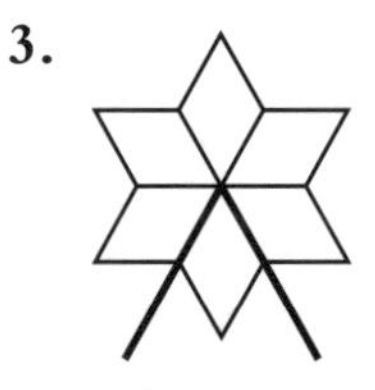

By moving the pattern block away from the mirrors and moving the mirrors closer to each other, it is possible to get seven copies of the block (including the original).

4, 5. 180°, 120°, 90°, 72°, 60°, 51.43°, 45°, 40°, 36°, 32.73°, 30°

6. One of the reflections of the mirror should be aligned with one of the mirrors (even cases); or two reflections should meet on the extension of the bisector of the angle between the two mirrors (odd cases).

7. Symmetric only: even cases
Symmetric or asymmetric: odd cases

8. Answers will vary.

Discussion Answers

A.

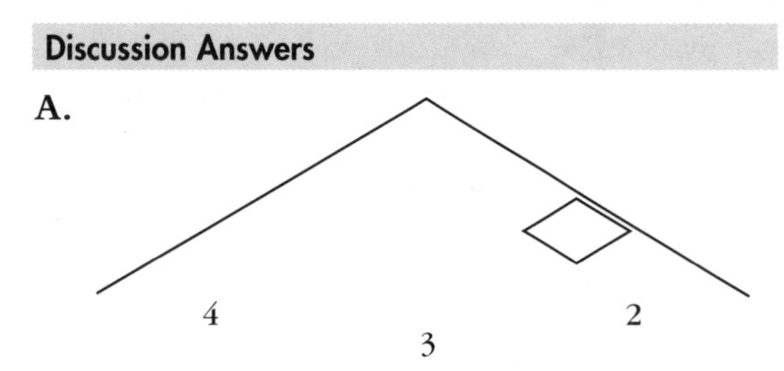

B. Mirrors: See answer to Problem 6. Objects: See answer to Problem 7.

C. (No question)

D. 120°–180°; 90°–120°; 72°–90°

E. The right eye of your reflection winks. This is unlike what happens in a mirror normally, because you are actually looking at the reflection of a reflection.

F. Every time you cross a mirror, you are looking at an additional level of reflection.

G. The reflection is backward, the reflection of the reflection is correct, and so on.

Lab 5.5: Rotation Symmetry

In chemistry, handedness of molecules is known as *chirality,* from the Latin root *chir-* (hand). This terminology in both languages comes from the fact that (in three dimensions) our hands are essentially identical, yet one cannot wear a left glove on a right hand. Hence the concept of handedness: It is about the difference between left- and right-handed figures.

You may also use Question G from Lab 5.4 (Two Mirrors) to discuss handedness.

You can interpret Problem 4 as a more open assignment, in which students can use any medium to create their designs. Possibilities include the following.

- any sort of geoboard
- pattern blocks
- any sort of dot or graph paper (polar graph paper works well)
- other shapes that can be found on the templates
- centimeter cubes
- tangrams
- compass and straightedge
- computer graphics programs
- geometry software
- potato prints
- rubber stamps
- anything!

If you decide to use potato prints or rubber stamps, be aware that unless the element you are reproducing is itself line symmetric you will need twin stamps that are mirror images of each other. Students can make rubber stamps in any shape at all with the materials sold by A Small Woodworking Company, 34207 82nd Ave. S., P.O. Box 460, Roy, WA 98580, (253) 458–3370. (They also sell pattern block rubber stamps.)

You can give more structure to the assignment by specifying values for n and specific numbers of each type of figure. You can increase motivation by promising that each student's best creations will be part of a bulletin board display.

You may complement Problem 4 by asking students to search for figures with the various sorts of symmetries in magazines or anywhere in the real world. Clippings, photographs, rubbings, or sketches of what they find can be included in the bulletin board exhibit. Things to look at include hubcaps, manhole covers, logos, and so on.

1. No answer is required.

2. a. 8 **b.** 3 **c.** 4 **d.** 6 **e.** 2 **f.** 12

3. a, 8; f, 12

4. Answers will vary.

Discussion Answers

A. b, c, d, e. Figures that do not have line symmetry are the ones that have handedness.

B. The number n must be a positive whole number. Any figure looks the same when rotated 360°, so any figure has one-fold symmetry. However, if that is the only symmetry a figure has, it is said to be not symmetric or *asymmetric*.

Lab 5.6: Rotation and Line Symmetry

The chart lists 72 possibilities, of which there are only 24 actual solutions. However, even 24 is a large number, and you should not expect each student to find and sketch all those solutions. Finding an example of each type could be a project for each group of four students or even for a whole class, with a checklist posted on the bulletin board along with colorful drawings of student-discovered designs. You should require students to label the designs according to their symmetry properties, and assign a couple of students to complete the exhibit by writing an explanation of rotation and line symmetry and the notation used in the table.

I have found that students are quick to find line-symmetric designs, but not so quick to find designs that have only rotation symmetry.

Answers

Answers will vary. See answer to Question A below.

Discussion Answers

A. Only one- and three-fold symmetries are possible for the triangle; one-, two-, three-, and six-fold symmetries are possible for the hexagon; all those, plus four- and 12-fold symmetries, are possible for the dodecagon. In each case, line symmetry is an option.

B. The number of sides of the figure is a multiple of n. The figure itself would not be able to be rotated by angles other than multiples of $360°/n$ and come back to its position.

C. Answers will vary. A combination of blocks can be replaced by other combinations that cover the same space. For example, two reds can replace a yellow.

D. Answers will vary.

E. Only one- or two-fold symmetries would be possible, since those are the only symmetries of the rhombus itself.

Lab 5.7: Two Intersecting Lines of Symmetry

You may demonstrate the given example, or another one, on the overhead to show the students that there are more reflections than one might expect at first. If they did Lab 5.4 (Two Mirrors), they should be ready for that idea.

Note that the intersecting-mirrors phenomenon that constitutes the subject of this lab is the underlying feature of kaleidoscopes. (A follow-up project could be to build a kaleidoscope.)

To save time and increase accuracy, you may provide assorted graph and dot papers, encouraging students to select carefully which sheet is most appropriate for each angle. You will find many special papers in the back of this book.

Question F does not need to be saved for last, as it contains a hint that will help speed up the process and make it more accurate: Additional lines of symmetry of the final figure can be obtained by reflecting the original ones.

Students can build on the sheets, using manipulatives of their (or your) choice. Then they can record the figure with the help of the template. Or they can use the template directly. You may hand out mirrors, hinged or not, as an aid in constructing and, especially, checking the designs. If you have Miras, those would help even more.

Of course, the list of manipulatives and other media suggested for creating rotationally symmetric designs applies to this lab as well. See the Notes to Lab 5.5 (Rotation Symmetry).

Note that Problem 3 is somewhat more difficult than Problem 2. Problems 3 and 4 could be skipped if time is short.

Answers

1. a.

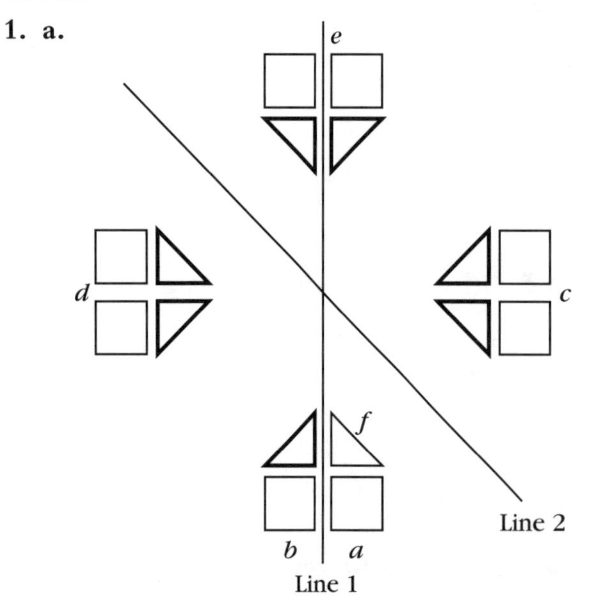

   b. Answers will vary.

2. Answers will vary.
3. Answers will vary.
4. 36° and 72°. If you reflect the mirrors themselves in the 72° case, you will end up with 36° between consecutive mirrors.

Discussion Answers

A. 90°, 4; 60°, 6; 45°, 8; 30°, 12; 72°, 10; 40°, 18; 36°, 10

B. In the even cases, the figures obtained have the same symmetry. In the odd cases, the figures are always symmetric. The difference is due to the fact that lines of symmetry are infinite in both directions, while our mirrors stopped where they met. Another difference is that lines of symmetry "work" in both directions, while real mirrors only work from one side. As a result, in this lab, unlike the one that used hinged mirrors, the direction you look from is irrelevant.

C. It is possible, if the object is placed in such a way as to have its own line of symmetry match the given line and if other identical objects are placed in the corresponding places as dictated by the symmetry.

D. In the case of the drawing, it is not necessary to place it in such a way as to have its own line of symmetry match the given line. After drawing the object, you can draw its reflection across the given line.

E. They are congruent and have the same handedness. One could be obtained from the other by a rotation around the point of intersection of the two lines of symmetry.

F. It makes sense. It does not change the final result, but it does make it easier to get there.

Lab 5.8: Parallel Lines of Symmetry

This lab works very much the same way as Lab 5.7 (Two Intersecting Lines of Symmetry), which is a bit of a prerequisite.

You may demonstrate the given example, or another one, on the overhead to show the students that the reflections go on forever. To save time and increase accuracy, you may draw the lines and duplicate them yourself. Or you can make available assorted graph and dot papers, encouraging students to select carefully which sheet is most appropriate for each angle.

Students can build on the sheets, using manipulatives of their (or your) choice. Then they can record the figure with the help of the template. Or they can use the template directly. You may hand out mirrors, hinged or not, as a help in constructing the designs. If you have Miras, those would help even more.

Of course, the list of manipulatives and other media suggested for creating rotationally symmetric designs applies to this lab as well. See the Notes to Lab 5.5 (Rotation Symmetry).

The symmetry of two parallel mirrors is often seen in designs such as the borders of rugs or architectural detail strips on buildings. If a third mirror is added at a special angle other than 90°, the symmetry becomes the symmetry of wallpaper designs, and the design expands to the whole plane.

The designs in Problems 5–7 are very difficult to generate because the pattern extends infinitely in all directions, wallpaper-style. You may want to assign those as extra credit and mount a bulletin board display of any nice creations.

Answers

1. Answers will vary.

2. Infinitely many

3–7. Answers will vary.

Discussion Answers

A. Parallel mirrors, like those in a barbershop, simulate parallel lines of symmetry.

B. They are congruent and have the same handedness. One could be obtained from the other by translation.

C. It makes sense. It does not change the final result, but it does make it easier to get there.

D. If the third mirror is perpendicular to the original two, the design stays confined to a strip in the direction of the perpendicular line. If the third mirror is not perpendicular or parallel to the first two, the design expands to cover the entire plane, like wallpaper.

E. In reflecting the mirrors, one obtains parallel lines.

6 Triangles and Quadrilaterals

Lab 6.1: Noncongruent Triangles

Prerequisites: Students should know the meaning of the word *congruent*.

This lab should precede or closely follow the introduction of congruent triangles. With a full discussion, it may well take two periods.

Students may have trouble understanding these instructions if they merely read them. Make sure you demonstrate Problem 1 (or Problem 2, which is more difficult) on the chalkboard or overhead.

Problem 7 is the famous "ambiguous case" and is quite difficult. You should probably demonstrate it at the board at the end of the lab, perhaps asking all students to copy your construction. It follows from this example that the fact that two triangles have an SSA correspondence does not mean they are congruent.

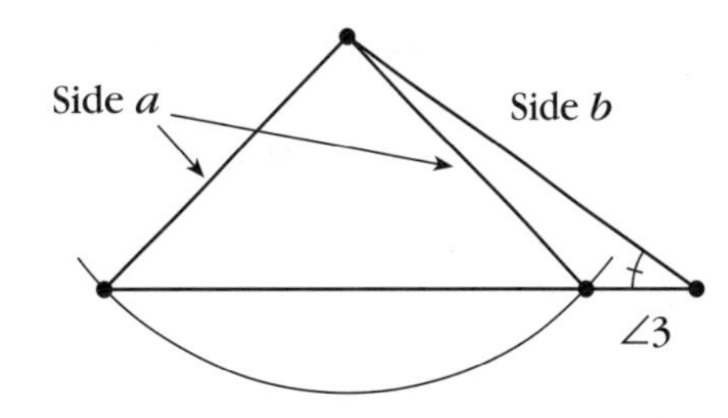

When discussing Question A, make sure students understand the difference between SAS and SSA.

Answers

1. Many triangles; SA

2. Many triangles; SA

3. Many triangles; AA

4. No triangles; the sum of angles is greater than 180°.
5. Exactly one triangle; SSS
6. Exactly one triangle; SAS
7. Exactly two triangles; SSA
8. Exactly one triangle; ASA
9. Exactly one triangle; SAA
10. No triangles; SSSA—too many constraints

Discussion Answers

A. See Notes and Answers above.

B. See Notes and Answers above. SSS, SAS, and ASA are criteria for congruence of triangles. So is SAA, which is essentially a version of ASA, because once two angles are determined, the third is too.

C. See Notes and Answers above.

D. Problem 7

Lab 6.2: Walking Parallelograms

This is an opportunity to review the angles created when a transversal cuts two parallel lines, while getting started on the general classification of quadrilaterals. Students often are confused about "Is a square a rectangle, or is a rectangle a square?" This lesson provides one way to get some clarity on this or at least provides an additional arena for discussion. See Lab 5.2 (Triangle and Quadrilateral Symmetry) for a complementary approach.

Students could be paired up, with each member of the pair executing the instructions written by the other member. The values of the variables must be stated before starting the walk.

In Problem 5, note that a variable should NOT be used in the case of the angles of the square and rectangle, since they are always 90°.

For Problems 9 and 10, you may give the hint that there are five true statements of the type "a rhombus is a parallelogram" that apply to the four quadrilaterals discussed.

To summarize, you may explain how a tree diagram or a Venn diagram can be used to display the relationships between the four figures.

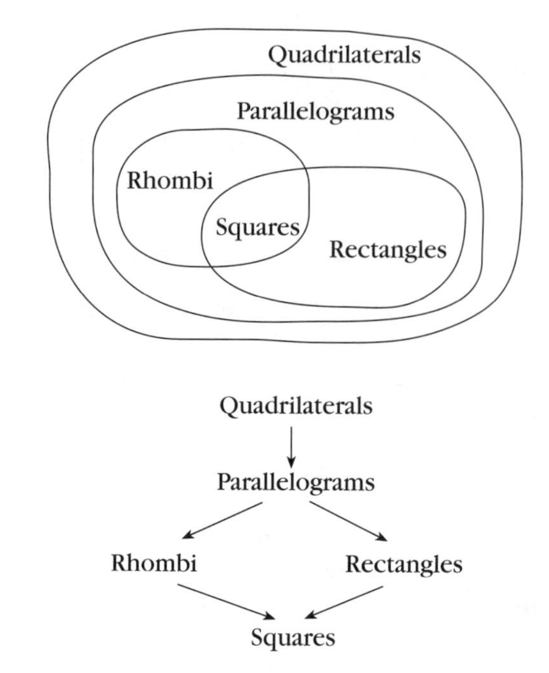

Question F is a way to follow up on a discussion about hierarchical relationships among quadrilaterals, rather than a direct extension of the lab. The answer to the question "Is a parallelogram a trapezoid?" depends on how a trapezoid is defined.

Answers

1. The first two turns add to 180°, since you're facing opposite to your original position.
2. Turn right 40°;
   Walk forward one step;
   Turn right 140°;
   Walk forward three steps;
   Turn right 40°.
3. 40° and 140°
4. Do this twice:
   Walk forward x steps;
   Turn right a°;
   Walk forward y steps;
   Turn right $(180 - a)$°.

5. Answers will vary. Here are some possibilities.

   a. Do this twice:

   Walk forward x steps;

   Turn right $a°$;

   Walk forward x steps;

   Turn right $(180 - a)°$.

   b. Do this twice:

   Walk forward x steps;

   Turn right 90°;

   Walk forward y steps;

   Turn right 90°.

   c. Do this four times:

   Walk forward x steps;

   Turn right 90°.

6. Yes. Use the side of the rhombus for both x and y.

7. No. It would be impossible to walk a parallelogram whose sides are of different lengths.

8. **a.** Yes.

   b. No.

9. Any square can be walked with parallelogram instructions (use equal sides and 90° angles), but not vice versa.

   Any square can be walked with rhombus instructions (use 90° angles), but not vice versa.

   Any square can be walked with rectangle instructions (use equal sides), but not vice versa.

   Any rectangle can be walked with parallelogram instructions (use 90° angles), but not vice versa.

10. A square is a parallelogram, a rhombus, and a rectangle.

    A rhombus and a rectangle are also parallelograms.

Discussion Answers

A. They are supplementary.
They are supplementary.

B. Parallelogram: three variables; rectangle and rhombus: two variables; square: one variable. The more symmetric a shape is, the fewer variables are used to define it.

C. Sides must be positive. Angles must be positive, but less than 180°.

D. Rectangle, rhombus

E. Rectangles: SS; rhombi: AS; parallelograms: SAS

F. If a trapezoid can have more than one pair of parallel sides, then parallelograms are particular cases of trapezoids. If not, there is no overlap between trapezoids and parallelograms.

Lab 6.3: Making Quadrilaterals from the Inside Out

This lab helps prepare students for the formal proofs of some quadrilateral properties. In addition to paper and pencils, you could have them use geoboards and rubber bands.

Another approach is to use (nonflexible) drinking straws or stirring sticks for the diagonals and elastic string for the quadrilateral. In the case of unequal diagonals, one of the straws can be cut. Straws can be connected with pins at the intersection. The elastic thread can be strung through holes punched near the ends of the straws or (easier but less accurate) a notch cut with scissors at the end of the straws.

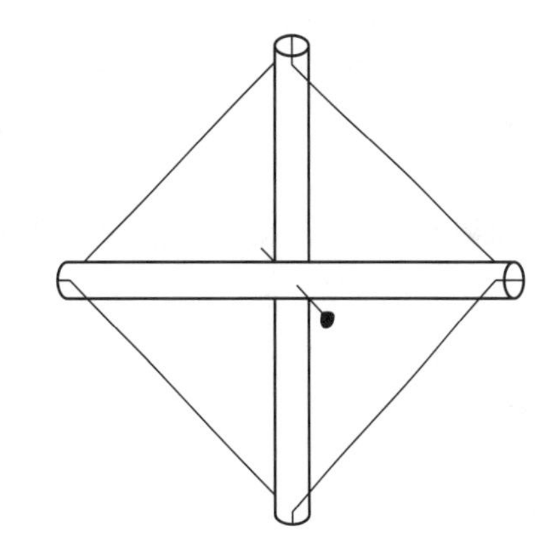

Make sure your students notice the suggestion that they try to make the most general quadrilateral that fits the given conditions. This is not easy, as students tend to be attracted to special, symmetric cases, especially when working on dot or graph paper.

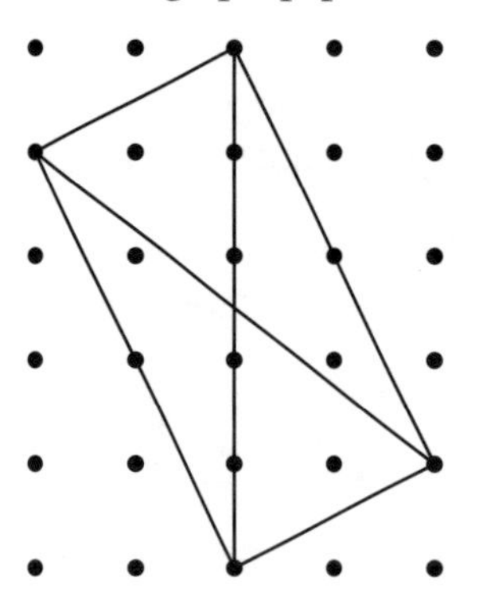

For Problems 5 and 6, one student used a 3, 4, 5 triangle on dot paper (with the legs horizontal and vertical) to get a diagonal of length 5 that is not perpendicular to a horizontal or vertical diagonal of length 5 (see figure). You may suggest this method to students who are stuck.

Answers

1. Square
2. General quadrilateral
3. Rhombus
4. General quadrilateral
5. Rectangle
6. General quadrilateral
7. Parallelogram
8. General quadrilateral

Discussion Answers

A. The diagonals do not bisect each other.

B. A kite requires perpendicular diagonals, one of which bisects the other, but not vice versa.

C. An isosceles trapezoid requires equal diagonals that intersect each other so that the corresponding segments formed are equal.

Lab 6.4: Making Quadrilaterals from Triangles

This assignment, like the previous one, helps prepare students for the formal proofs of some quadrilateral properties. You may use it as a homework assignment.

In any case, precede it with a discussion of how to present the information. Some students have done this in a table format, others in paragraphs, and other ideas are certainly possible. Emphasize that it is essential to clearly illustrate the report.

Problem 2 is not crucial, but it allows trapezoids to come into the activity.

Answers

1. EQ: rhombus
   AI: kite, parallelogram, rhombus
   OI: kite, parallelogram, rhombus
   AS: kite, parallelogram
   RS: kite, rectangle, parallelogram
   HE: kite, rectangle, parallelogram
   OS: kite, parallelogram
2. In addition to the quadrilaterals that can be made with two triangles, it's possible to make trapezoids, isosceles trapezoids, and right trapezoids.

Lab 6.5: Slicing a Cube

This activity is highly popular with students. The discoveries of the equilateral triangle and, especially the regular hexagon are quite thrilling. Plan on at least two class periods to do it all. As a way to get started, students may use the template to draw various shapes on the stiff cardboard and test whether those can be cube slices.

It is possible to follow up this lab with an exhibit featuring 2- and 3-D illustrations. Here are some of the possibilities I suggested to my students (they did the work as homework).

A "movie" of the cube moving through a plane in various ways:

- face first
- edge first
- vertex first

Alternatively, the cube could be rotating around an axis and passing through a plane containing that axis. The axis could be:

- an edge
- a diagonal
- some other line

)r students could analyze:

- possible and impossible triangle slices
- possible and impossible quadrilateral slices
- possible and impossible pentagon and hexagon slices

n all cases, students should pay attention to:

- the ranges of possible side lengths
- the ranges of possible angles

tudents should pay particular attention to pecial figures:

- the ones where a change occurs (from triangle to trapezoid, for example)
- the most symmetric ones
- the largest ones

The projects should include some text and ome 2- or 3-D visual components. When ompleted, they can be put together in one nuseum-style exhibit.

ome of my students made clay models, others sed the transparency cube, and others drew -D sketches. A couple of them chose to write xplanations of why right and obtuse triangles re impossible. Two commented on the fact that ome slices appear possible when drawn in two limensions, but then turn out to be impossible o execute in three dimensions. If your students re familiar with the Pythagorean theorem and he basic trig ratios, the exhibit could have a uantitative component, with calculations of he sides of certain figures.

nswers

, 2. No answer

3. Triangles:

Equilateral (EQ): possible
Acute isosceles (AI): possible
Right isosceles (RI): impossible
Obtuse isosceles (OI): impossible
Acute scalene (AS): possible
Right scalene (RS): impossible
Half-equilateral (HE): impossible
Obtuse scalene (OS): impossible

Quadrilaterals:

Square (SQ): possible
Rhombus (RH): possible
Rectangle (RE): possible
Parallelogram (PA): possible
Kite (KI): impossible
Isosceles trapezoid (IT): possible
General trapezoid (GT): possible
General quadrilateral (GQ): impossible

Other polygons:

Pentagon: possible
Regular pentagon: impossible
Hexagon: possible
Regular hexagon: possible
Seven-gon: impossible

Discussion Answers

A. See above.

B. If the edge of the original cube is 1, the length of the sides of the slices is anywhere from 0 to $\sqrt{2}$. The angles are from 0° to 90°.

C. An equilateral triangle, a square, and a regular hexagon are all possible. A regular pentagon is not possible, though a pentagon with one line of symmetry is possible.

D. Yes. Start with a square slice sitting at the bottom of the cube. As you slide one side of the slice up a vertical face of the cube, the opposite side will slide along the bottom. The square will still slice the cube, but it won't be parallel to any face.

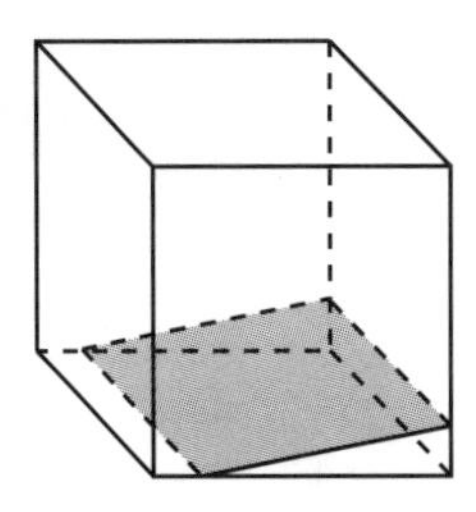

7 Tiling

Lab 7.1: Tiling with Polyominoes

Prerequisites: Students need to be familiar with the polyomino names. Since many students will have forgotten the polyomino vocabulary (which is not particularly important to remember), you may remind them of (or supply them with a copy of) the Polyomino Names Reference Sheet (page 54).

Question B is quite difficult. One way to think about it is to break it down into two separate dimensions: First analyze the strips in the design and establish that those strips can be extended indefinitely, then judge whether the strips can be placed next to each other without holes or overlaps. Of course, some tilings may not lend themselves to this analysis. In fact, that in itself is an interesting question: Can you devise a polyomino tiling that seems to work yet does not consist of strips?

Question D applies a concept that was introduced in Lab 4.2 (Polyominoes and Symmetry).

Extension: Is it possible to tile the plane with each of the 35 hexominoes?

Answers

1. Answers will vary.
2. No.

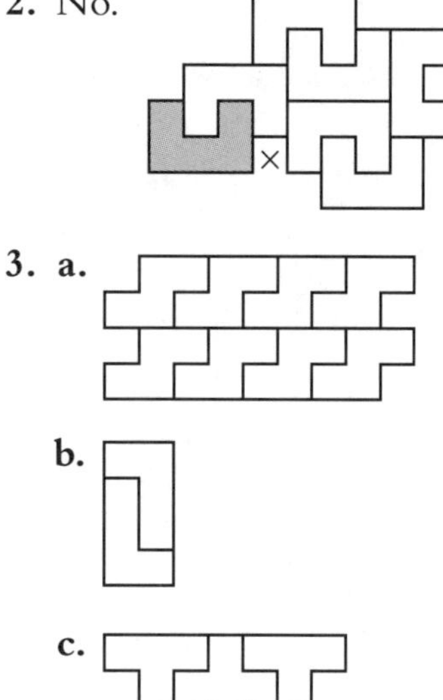

3. a.

b.

c.

4.

Discussion Answers

A. Yes.

B. See Notes above.

C. Yes. See answers to Problems 3 and 4.

D. Answers will vary.

E. Answers will vary.

F. Answers will vary.

Lab 7.2: Tiling with Pattern Blocks

This lab follows up the early pattern block labs on angles (see Lab 1.1: Angles Around a Point) and helps lay the groundwork for Lab 7.3 (Tiling with Triangles and Quadrilaterals). While this lab is not a prerequisite to Lab 7.3, it does bring students' attention to the sum of the angles around a vertex. This is the key mathematical concept here, and every student should be absolutely clear on this.

or Problem 2, make sure the students nderstand that they are to color the ertices, not the polygons.

Question B can lead in a couple of different irections: For example, one could avoid repetition y writing 4·o90 for the vertex in a checkerboard attern; or one could reduce the number of colors nentioned or eliminate the mention of colors together (although the latter approach would robably lead to ambiguous names). Another ossibility is to omit the numbers in the case f the triangle, square, and hexagon, since the umber is unique.

Discussions of notation are a common part f mathematics and are a legitimate secondary chool activity; to avoid such discussions bscures the fact that mathematical notations re not "natural" but rather are created over me. A notation becomes standard only if nough mathematicians accept it.

nswers

1. Answers will vary.
2. The first kind of vertex is where only triangles meet. The other one is any other vertex.
3. 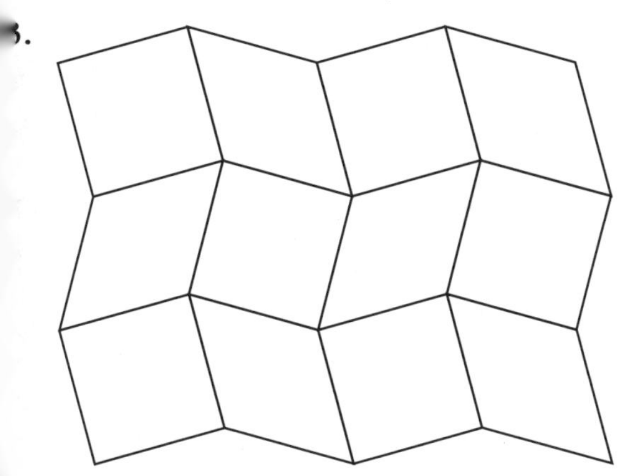
4. Answers will vary.
5. 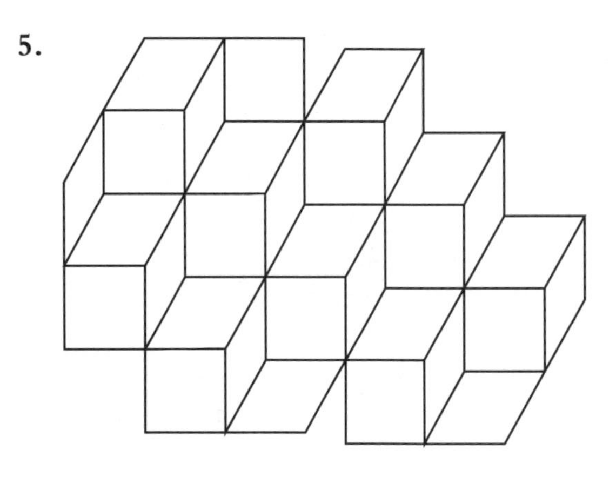
6. Answers will vary.
7. The angles do not add up to 360°.

Discussion Answers

A. They must add up to 360°. An arrangement of pattern blocks around a vertex can be the start of a tiling pattern.

B. See Notes above.

C. Answers will vary.

Lab 7.3: Tiling with Triangles and Quadrilaterals

In Problem 1, do not expect each student to discover all 16 tilings alone—it would be time-consuming and repetitive. Groups of two to four students can work on it, and the final results can be colored and displayed on the bulletin board. If you are short on time, only ask for a scalene triangle and general quadrilateral tilings. If students come up with shortcuts based on observations like the ones in Question B, that will speed up the process as well as develop their visual experiences with triangles and quadrilaterals.

Students may have trouble with some cases, however, particularly the asymmetrical ones that are addressed in Problems 2–4. If so, you may give some or all of the following hints.

- Match like sides.
- Make strips, then juxtapose them.
- It is not necessary to flip the figures over.
- All the angles around a vertex must add up to 360°. How can this be accomplished given

what you know about the angles of a triangle? About the angles of a quadrilateral?

The final point is probably the key.

One technique that facilitates exploration is to take a sheet of paper, fold it in half three times, then draw a triangle or quadrilateral on it. If you cut it out, you get eight identical copies, and that's usually enough to conduct experiments. Be careful, though, that you don't turn over any of the triangles or quadrilaterals. To avoid that problem, make sure all the copies are placed on the table with the same face up, and mark or color that face.

The explanations in Problems 2 and 4 are not easy to write and may take you beyond one class period. You may ask students to write only one of the two. In any case, encourage your students to discuss their approaches with each other and perhaps with you and to write drafts before embarking on the final version. Hopefully, the points listed in the hints above will appear in the students' write-ups. Remind them that well-labeled illustrations are a must in a report of this sort.

Answers

1. Answers will vary.
2. No. Possible explanation: Any two copies of a triangle can be arranged to make a parallelogram. The parallelograms can be arranged to make strips.
3. Answers will vary.
4. Copies of the quadrilaterals must be arranged so that matching sides are adjacent and all four angles are represented at each vertex. An example is shown here.

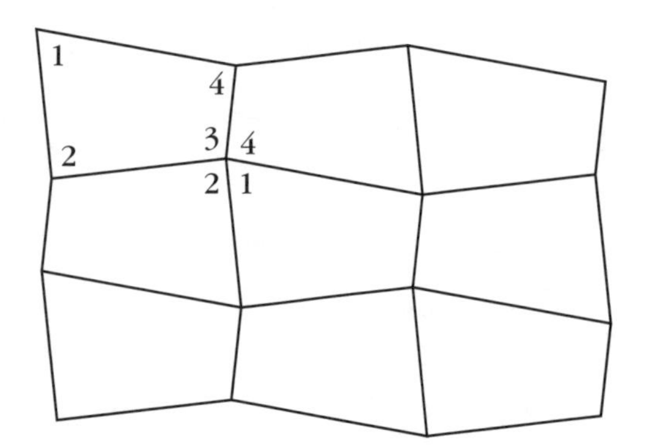

Discussion Answers

A. Answers will vary.

B. Answers will vary.

C. Answers will vary.

Lab 7.4: Tiling with Regular Polygons

Problem 1 revisits work students may have already done in Lab 7.2 (Tiling with Pattern Blocks). In fact, for some of the regular polygons it is convenient to use the pattern blocks. For others, see the Note in Lab 7.3 (Tiling with Triangles and Quadrilaterals) about cutting out eight copies of the tiles in order to experiment.

If students are stuck on Problem 2, you may suggest as a hint that they read the text following Problem 3.

Problem 4 is yet another take on the "angles around a point" concept. See Labs 1.1 (Angles Around a Point), 7.2, and 7.3. Note that an arrangement of regular polygons around a point is a necessary but not sufficient condition for a tiling.

In Question C, eight of the eleven tilings can be made with pattern blocks, and all can be made with the template.

If some students are particularly interested in these questions, you could ask about the possibility of tiling with regular polygons that are not on the template. It is not too difficult to dismiss 9- and 11-gons by numerical arguments based on their angles.

Answers

1. Equilateral triangle, square, regular hexagon
2. Their angles are not factors of 360°.
3. Answers will vary.
4. There are twelve ways to do this, including three one-polygon solutions, six two-polygon solutions, and three three-polygon solutions:

   6 triangles
   4 squares
   3 hexagons

2 hexagons, 2 triangles
1 hexagon, 4 triangles
2 octagons, 1 square
2 dodecagons, 1 triangle
2 squares, 3 triangles
1 decagon, 2 pentagons

1 hexagon, 2 squares, 1 triangle
1 dodecagon, 1 hexagon, 1 square
1 dodecagon, 1 square, 2 triangles

5. Answers will vary.

6. Answers will vary. Example: If you start a tiling by surrounding a vertex with two pentagons and a decagon, you find that a third pentagon is necessary next to the first two. On the other side of the pentagons, you are forced to use two decagons, but those must overlap each other, which ruins the tiling.

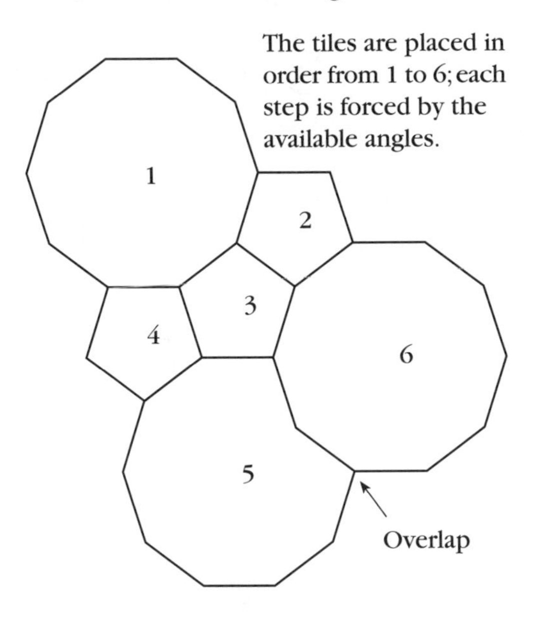
The tiles are placed in order from 1 to 6; each step is forced by the available angles.
1
2
3
4
5
6
Overlap

Discussion Answers

A. This can be done with equilateral triangles; squares; and hexagons with triangles.

B. 15-gon: triangle, decagon
18-gon: triangle, nonagon (9-gon)
20-gon: square, pentagon
24-gon: triangle, octagon
42-gon: triangle, heptagon

C.

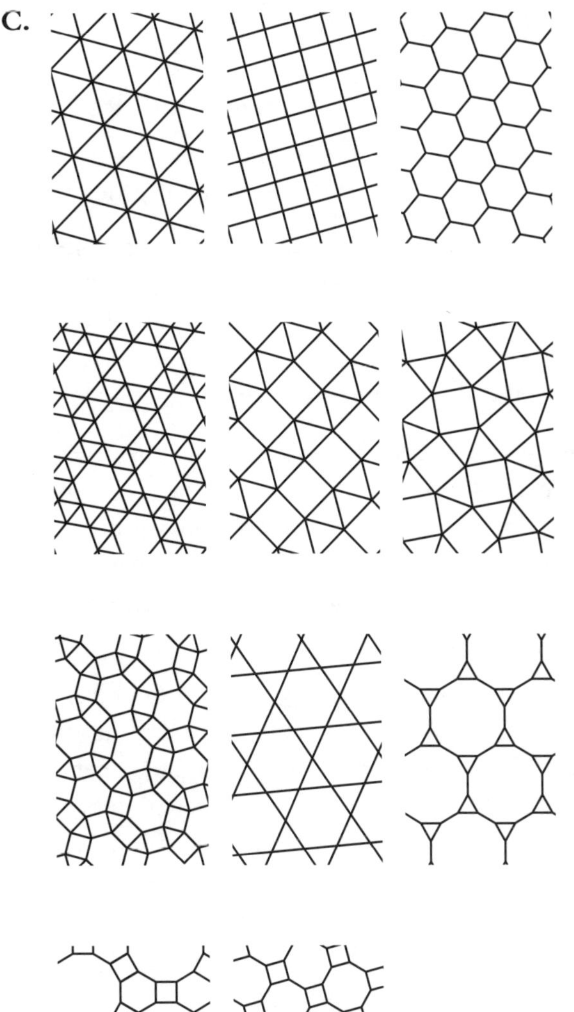

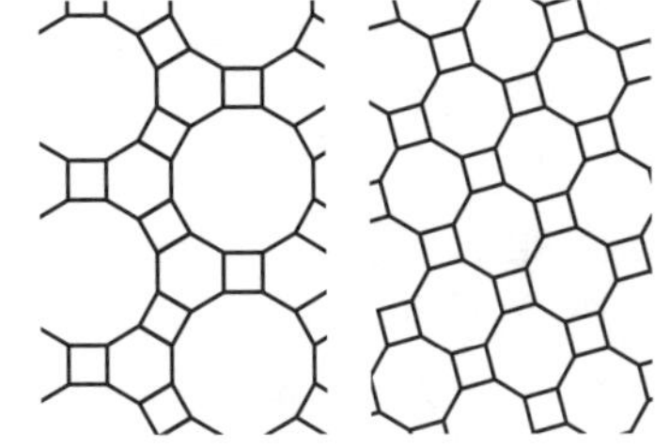

D. Answers will vary. Most tilings will have parallel and nonparallel lines of symmetry. Nonparallel lines will form 30°, 45°, 60°, or 90° angles. Centers of rotation symmetry will be found at centers of polygons, at midpoints of edges, and at vertices. They may have the same or different *n*-fold symmetries: two-fold, three-fold, four-fold, or six-fold.

8 Perimeter and Area

Lab 8.1: Polyomino Perimeter and Area

This lab is highly recommended, as it is usually guaranteed to generate intense involvement.

Prerequisites: It is not necessary to have done the labs in Section 4 (Polyominoes) in order to do this lab. The definition of polyominoes given here is equivalent to the one from Section 4 and is more relevant to this section.

Timing: This can easily be a two-day investigation, especially if you take the time to have students make the graph as suggested in Question C, then discuss the three-way relationship between the figures, the numbers in the table, and the graph.

Problems 1 and 2 remind students of the meaning of perimeter and area and establish the basic questions of the investigation. Filling out the table can seem overwhelming, but a pattern quickly emerges for the longest perimeters. The shortest perimeters are more difficult to understand, but they usually generate a great deal of curiosity (see below). The prediction exercises are the students' opportunity to articulate the patterns they discovered while making the table.

After some false starts, students usually discover that the shortest perimeters increase by 2 in the following manner: one 4, one 6; two 8s, two 10s; three 12s, three 14s; and so on. On the graph, this shows up as a staircase whose steps get longer at every other step. More patterns will be discussed in the following lab.

One way to answer Question D is to use the graph. For example, to figure out whether there is a polyomino with both area and perimeter equal to 14, one can plot the point (14, 14) and observe whether it is in the feasible area (the area that lies between the two graphs). The line $P = A$ can be graphed on the same axes, which will make it clear that there are polyominoes with equal area and perimeter for all even numbers greater than or equal to 16.

Answers

1. H: area 7, perimeter 16
   P: area 7, perimeter 12
2. Answers will vary.
3.

Area	Minimum perimeter	Maximum perimeter
1	4	4
2	6	6
3	8	8
4	8	10
5	10	12
6	10	14
7	12	16
8	12	18
9	12	20
10	14	22
11	14	24
12	14	26
13	16	28
14	16	30
15	16	32
16	16	34
17	18	36
18	18	38
19	18	40
20	18	42
21	20	44
22	20	46
23	20	48
24	20	50
25	20	52
26	22	54

4. $P_{max} = 2A + 2$

5. Answers will vary. See Labs 8.2 and 8.3.

6. a. 49 min. 28 max. 100

b. 45 min. 28 max. 92

c. 50 min. 30 max. 102

d. 56 min. 30 max. 114

Discussion Answers

A. No. Every unit of length in the perimeter is matched by another unit on the opposite side, so the total must be even.

B. Answers will vary.

C. The maximum perimeter is a linear function of the area. The minimum is a step function, with the steps getting longer and longer. See graph on next page.

D. The smallest such polyomino is a 4×4 square, with area and perimeter equal to 16. All even areas greater than 16 can be arranged to have a perimeter equal to the area. See Notes.

E. Answers will vary. See Lab 8.2.

Lab 8.2: Minimizing Perimeter

Prerequisite: Lab 8.1 (Polyomino Perimeter and Area) is a prerequisite, since this lab attempts to develop a geometric insight into the toughest part of that lab, the minimum perimeter for a given area.

Finding a formula for the minimum perimeter is too difficult for most students, but the insights generated in this lesson can lead to the following algorithm.

- Find out whether A is a square number, that is, $A = n^2$ for some n, or a rectangular number, that is, $A = n(n + 1)$ for some n. If so, the minimum perimeter is $4n$ in the first case and $4n + 2$ in the second case.
- If A is not a square number, then the minimum perimeter is the same as that for the next higher number that is a square or rectangular number.

Actually, there is a formula for the minimum perimeter. It is

$$P = 2\lceil 2\sqrt{A} \rceil,$$

where $\lceil\ \rceil$ represents the ceiling function. The ceiling of a number is the least integer greater than or equal to the number. For example:

$$\lceil 2 \rceil = 2$$

$$\lceil 2.1 \rceil = 3$$

$$\lceil 2.8 \rceil = 3$$

Explanation: It is easy to see that, in the case of area A as a perfect square, the formula for the minimum perimeter can be written $P = 4\sqrt{A}$, since $\sqrt{A}$ gives you the side of the square. In the case of an area less than a perfect square but greater than the immediately preceding rectangular number, the formula becomes $P = \lceil 4\sqrt{A} \rceil$. The difficulty comes in the case of areas equal to or immediately less than a rectangular number. A little experimentation will show you that, in this case, the formula given above works in all cases. The proof is based on the above insights, themselves a consequence of the spiral argument suggested at the end of the lab. It is presented here for your benefit, as it is unlikely to be meaningful to your students.

Proof: There are two cases, as described above.

First case: If $(n - 1)n < A \leq n^2$, where n is a natural number, then $P = 4n$. By distributing, the compound inequality can be rewritten: $n^2 - n < A \leq n^2$. But since we are only talking about whole numbers, it is still true (via completing the square) that $n^2 - n + \frac{1}{4} < A \leq n^2$, and therefore

$$\left(n - \frac{1}{2}\right)^2 < A \leq n^2 \quad \text{Factor.}$$

$$n - \frac{1}{2} < \sqrt{A} \leq n \quad \text{Square root.}$$

$$2n - 1 < 2\sqrt{A} \leq 2n \quad \text{Multiply by 2.}$$

(Proof continues on page 211.)

8.1 C

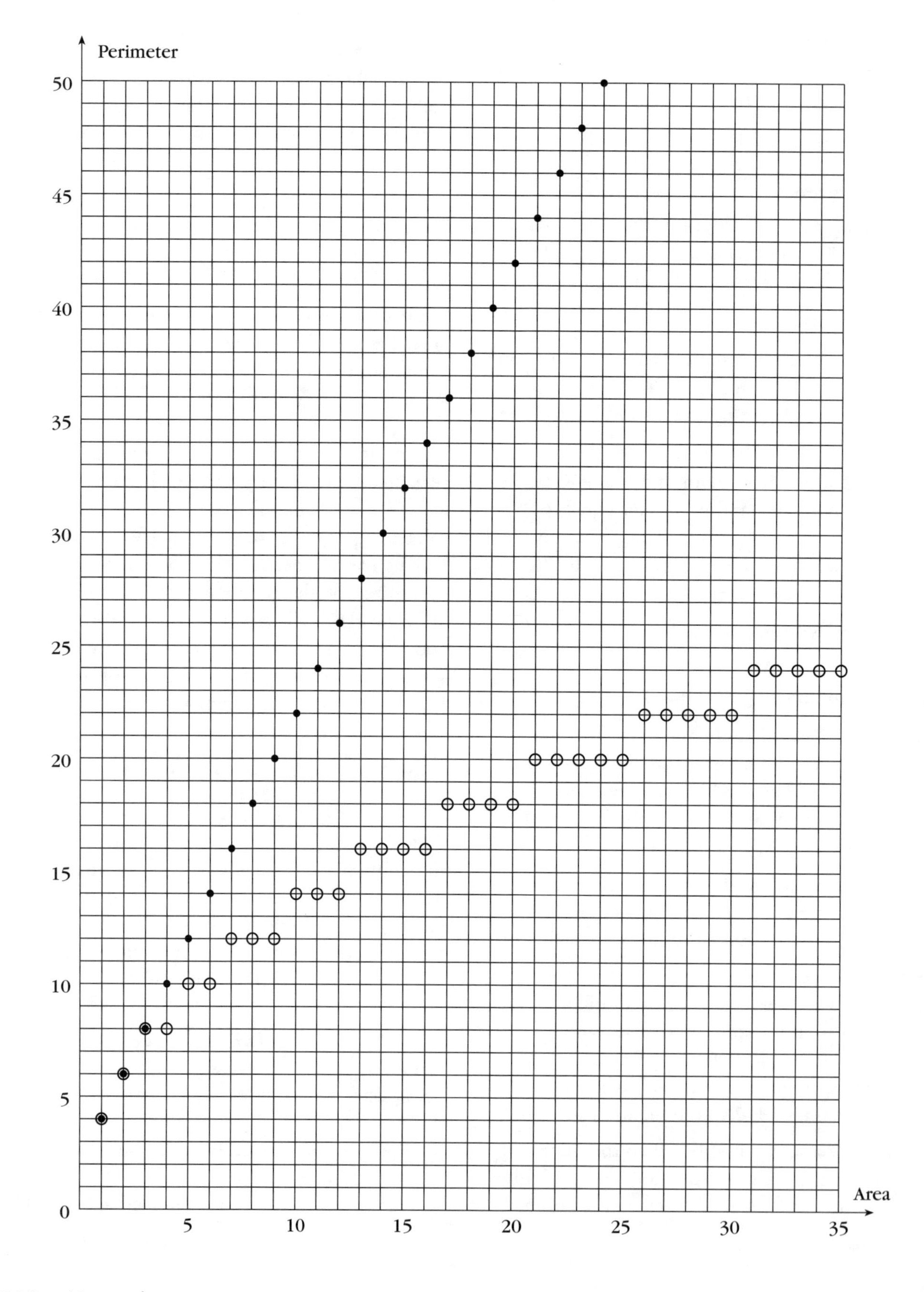

This shows that $\lceil 2\sqrt{A} \rceil = 2n$, and since $P = 4n$, the equation for minimum perimeter is $P = 2\lceil 2\sqrt{A} \rceil$.

Second case: If $n^2 < A \le n(n + 1)$, then $P = 4n + 2$. A similar argument to the previous one leads to the inequality

$$2n < 2\sqrt{A} \le 2n + 1.$$

In this case, $\lceil 2\sqrt{A} \rceil = 2n + 1$, and since $P = 4n + 2$, once again the equation for minimum perimeter is $P = 2\lceil 2\sqrt{A} \rceil$.

Answers

1. They alternate between two types of special numbers: perfect squares and the products of consecutive whole numbers.
2. 1, 4, 9, 16, 25
3. 2, 6, 12, 20
4. **a.** 120
   b. 122
   c. 120
5. $29 \cdot 30 = 870$, $30 \cdot 31 = 930$
6. **a.** 122
   b. 124
   c. 118
7. The perimeter increases by 2 whenever a unit square is added to a perfect square or rectangle.

Discussion Answers

A, B. Squares (and rectangles that are nearly square) are compact and minimize perimeter. In building the spiral, adding a square usually covers two side units from the old perimeter and adds two new side units, which is why the perimeter does not increase. However, when you add a piece to a square or rectangle, you are adding 3 and subtracting 1, so the net effect is to increase the perimeter by 2.

C. a. $4n$

b. $4n + 2$

Lab 8.3: A Formula for Polyomino Perimeter

The formula relating area to perimeter and inside dots is a particular case of Pick's formula, which will be explored in its full generality in Lab 8.6 (Pick's Formula).

The puzzle in Problem 7 is not closely related to the rest of the page, but it can be useful in keeping your fastest students occupied while their classmates continue to work on the previous questions.

The wording of Question D is slightly misleading in that there actually is no greatest area for a given number of inside dots, since one can append an inside-dotless tail of any length to a figure.

Questions C and D suggest the related questions: What is the greatest number of inside dots for a given area? What is the least area for a given number of inside dots? These questions are difficult because they are related to the problem of the minimum perimeter for a given area.

Answers

1.

Perimeter	Area	Inside dots
12	9	4
14	9	3
18	11	3

2. Answers will vary.
3. You are subtracting 2 from the perimeter.
4. Answers will vary.
5. You are adding 2 to the perimeter.
6. $P = 2A - 2i + 2$
7. 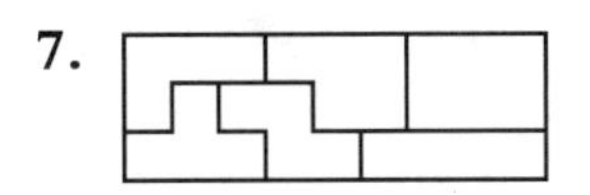

Discussion Answers

A. It is the same.

B. The first one has fewer inside dots.

C. 0

D. There is no greatest area. See Notes for this lab.

E. There are $(n - 1)^2$ inside dots.

F. There are $n(n - 1)$ inside dots.

Lab 8.4: Geoboard Area

This lab sets the foundation for later work, especially in Section 9 (Distance and Square Root).

The opening problem is deliberately open-ended. It provides a friendly entry into the subject: Students will find their own level, as there are not only plenty of easy solutions, but also many very tricky ones. If students seem stuck on rectangles, praise and share more eccentric discoveries. A discussion of student creations on the overhead can serve as a foundation for the whole lab. You may also consider going back to Problem 1 to seek additional solutions after doing the rest of the lab.

The best approach for Problem 2 is to build a rectangle around the triangle and divide its area by 2.

Students often have difficulty thinking of the subtraction method that is necessary for Problems 4 and 5. It may be useful to discuss an example or two on the overhead. This technique will be important later on, so make sure all the students understand it.

On the other hand, the last two problems are not essential. Problem 6 is an opportunity to discuss the shearing of triangles, which can be useful in thinking about complicated geoboard areas. Problem 7 is an opportunity to practice what was learned in this lab.

Answers

1. Answers will vary.
2. 1.5; 3
3. 6; 6; 4.5
4. 2.5
5. 4; 5; 6
6. Base 8, height 2: seven triangles

   Base 4, height 4: nine triangles

   Base 2, height 8: ten triangles (including one that is congruent to a base-8, height-2 triangle)

   So there is a total of 25 noncongruent triangles with an area of 8.
7. 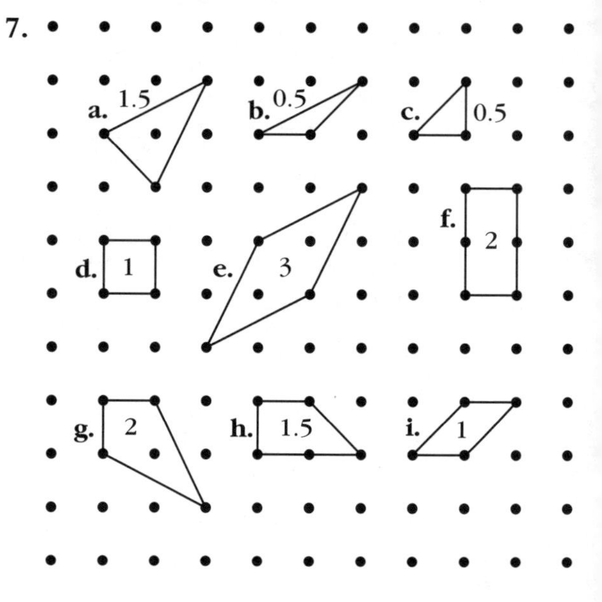

Discussion Answers

A. The mistake results from counting the pegs rather than finding the lengths of the sides.

B. Answers will vary but should include the following ideas.

- Division by 2: "Easy" right triangles are half of a rectangle.
- Addition: Complicated figures can sometimes be broken up into smaller ones.
- Subtraction: Complicated figures can be surrounded by a large rectangle or square.

C. The area remains constant, and equal to half the product of base and height.

D. Answers will vary.

Lab 8.5: Geoboard Squares

Prerequisites: In order to find the areas of the squares, students should have done Lab 8.4 (Geoboard Area), where the necessary techniques are introduced.

Timing: Even though the main part of the lab has only one question, it will take more than one period. It's worth it, because this is a crucially important lab that develops students' visual sense, prepares them for the work on distance in Section 9, and leads to a proof of the Pythagorean theorem.

At first, students only find the horizontal-vertical squares. Then they usually discover the squares that are tilted at a 45° angle. As soon as someone finds another square past the first 15, I usually hold it up for everyone to see, as a hint about how to proceed.

For many students, it is extraordinarily difficult to actually find the tilted squares. One approach, which helps answer Questions A and B, is to make an "easy" square, say, 5 × 5. Then the vertices of another square can be obtained by moving clockwise from each vertex a given number of units. The resulting figure is perfectly set up for finding the area.

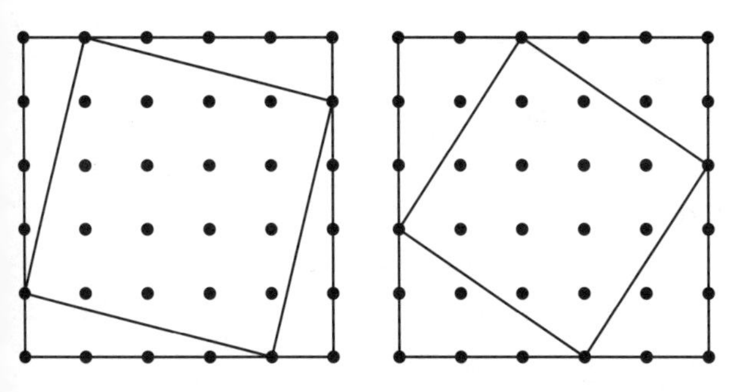

Students may find what appear to be 35 different squares, but once they calculate the areas they will see that there are duplicates.

If your students are not yet comfortable with the concept of square root, this lab is a perfect opportunity to talk about it. The horizontal-vertical squares provide an easy way to start the discussion: The side of a square of area 49 is 7, or $\sqrt{49}$. In the case of the tilted squares, if the area is 5, then the side must be $\sqrt{5}$.

Make sure students save the results of their work, as it will be useful when working on Lab 9.1 (Taxicab Versus Euclidean Distance).

Note that Question D outlines a standard proof of the Pythagorean theorem. If your students have studied congruent triangles, the proof can be made rigorous by first proving that the initial shape is indeed a square. The fact that they will have drawn the figures and done these calculations repeatedly when working on Problem 1 will make the proof accessible to many more students.

Answers

1. Areas: 1, 2, 4, 5, 8, 9, 10, 13, 16, 17, 18, 20, 25, 26, 29, 32, 34, 36, 37, 40, 41, 45, 49, 50, 52, 53, 58, 64, 65, 68, 81, 82, 100. Sides: the square roots of the areas.

Discussion Answers

A. Measure with the corner of a piece of paper. Alternatively, count pegs up and across for a given side, and check that by rotating the board 90° you still have the same counts for the other sides.

B. One method is to start with each of the "easy" squares, using the method outlined in the Notes to this lab to find all the squares that are nested within it.

C. Yes, there are squares with area 25 in two different orientations and squares with area 50 in two different orientations.

D. **a.** $a + b$

b. $(a + b)^2$

c. $\frac{ab}{2}$

d. $(a + b)^2 - 4\left(\frac{ab}{2}\right) = a^2 + b^2$

e. $\sqrt{a^2 + b^2}$

Lab 8.6: Pick's Formula

Prerequisites: Lab 8.4 (Geoboard Area) provides the necessary foundation. This lab is also related to Lab 8.3 (A Formula for Polyomino Perimeter), although that is a prerequisite only for Question C.

Pick's formula is a surprising result, and the search for it is a worthwhile mathematical challenge at this level. There is no need for students to memorize it, but if they do, they may find it useful when working on Lab 9.1 (Taxicab Versus Euclidean Distance).

Answers

1. 4.5
2.

Inside dots	Boundary dots	Area
3	5	4.5
4	4	5
2	12	7

3. Answers will vary.
4. The area increases by 1.
5. Answers will vary.
6. The area increases by $\frac{1}{2}$.
7. $A = i + \frac{b}{2} - 1$

Discussion Answers

A. There is no limit to the area for a given number of inside dots, because a tail of any length can be added to the figure without adding any more inside dots.

B. $A = \frac{b}{2} - 1$

C. It is essentially the same formula, though it is more general. The perimeter of a polyomino is equal to its number of boundary dots.

9 Distance and Square Root

Lab 9.1: Taxicab Versus Euclidean Distance

Prerequisites: Lab 8.5 (Geoboard Squares) is essential, both for its introductory work with square roots and for laying the visual and computational groundwork for the calculation of Euclidean distance.

The purpose of Problem 1 is to clarify something that Euclidean distance is not and also to help frame students' thinking about Euclidean distance, since in both cases it is useful to use horizontal and vertical distance (run and rise, as we say in another context). If you think students need to work more examples, those are easy enough to make up. In fact, you may ask students to make up some examples for each other.

For Problem 2, students can use either their records from Lab 8.5 or Pick's formula to find the area of the squares, or the Pythagorean theorem if they think of it and know how to use it.

Problem 3 is an interesting but optional extension. For a discussion of a taxicab geometry problem related to Problem 3, see Lab 9.6 (Taxicab Geometry).

Question A requires the use of absolute value. If your students are not familiar with it, you may use this opportunity to introduce the concept and notation, or you may skip this problem.

The answer to Question C is a consequence of the triangle inequality (in Euclidean geometry).

We return to taxi-circles in Lab 9.6.

Answers

1. **a.** 11
   b. 10
   c. 4
   d. 6.4
   e. 3.24
   f. 5.64
2. **a.** $\sqrt{65}$
   b. $\sqrt{29}$
   c. 2.5
   d. 5
   e. 0.9
3.

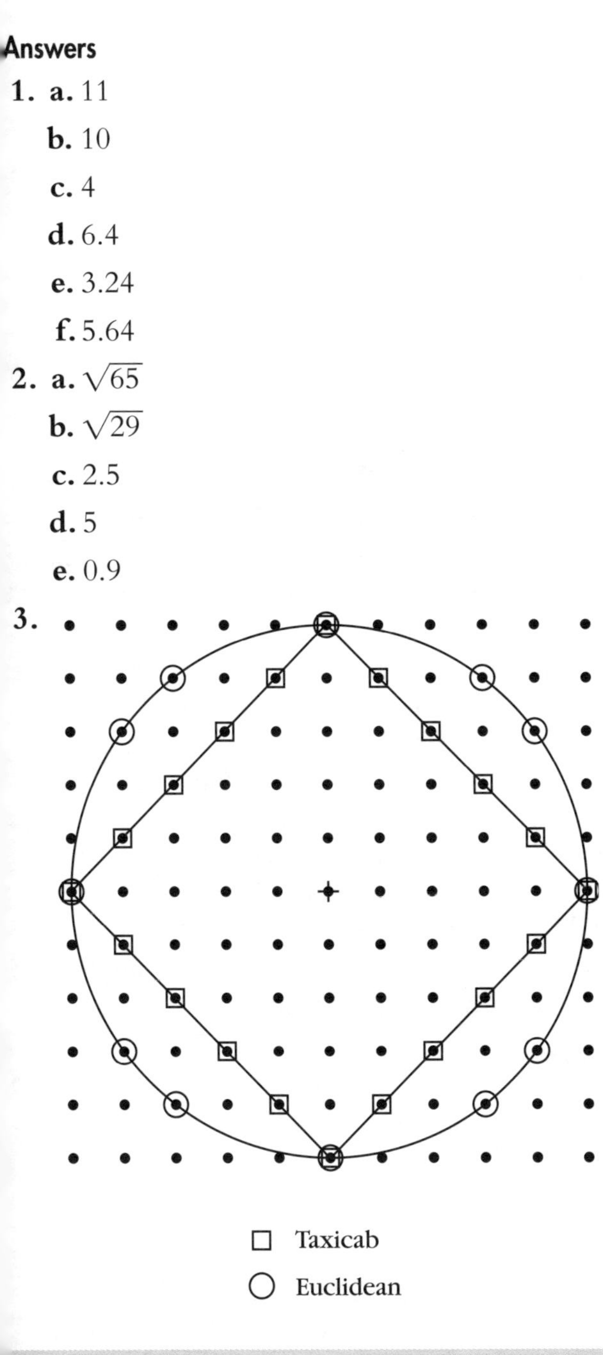

Discussion Answers

A. $T(P_1, P_2) = |x_1 - x_2| + |y_1 - y_2|$

B. Yes. We have equality if B is inside the horizontal-vertical rectangle for which A and C are opposite vertices.

C. Taxicab distance is greater if the two points are not on the same horizontal or vertical line. If they are, taxicab and Euclidean distances are equal.

D. Because a circle is a set of points at a given distance from a center

Lab 9.2: The Pythagorean Theorem

Prerequisites: This lab wraps up a series that started with Lab 8.4 (Geoboard Area). The most important labs in the series are 8.4, 8.5, and 9.1.

Dot paper will work better than geoboards, because the figures are difficult to fit in an 11 × 11 board. However, there is no reason to prevent students who want to work on the geoboards from doing that.

Of course, the experiment in Problem 2 does not constitute a proof of the Pythagorean theorem. In Lab 8.5, Question D outlines the key steps of such a proof. Still, it's important to go through this activity because many students memorize the Pythagorean theorem (which is certainly among the most important results in secondary math) with little or no understanding. This lab, particularly if it is followed by Question A, helps make the theorem more meaningful.

You may ask students to memorize the formula $a^2 + b^2 = c^2$ or, perhaps, the less traditional but more explicit $\text{leg}_1{}^2 + \text{leg}_2{}^2 = \text{hyp}^2$.

If you don't have time for it, it is not necessary to get a complete list as the answer to Question C; a few examples are enough. If your students are interested in getting a full listing, it will be easy to obtain after doing Lab 9.4 (Distance from the Origin).

Answers

1.

Area of squares		
Small	**Medium**	**Large**
5	20	25

2. Answers will vary.
3. Area of small square + area of medium square = area of large square

4. In a right triangle, the sum of the areas of the squares on the legs equals the area of the square on the hypotenuse.

5. **a.** $\sqrt{61}$

   b. $\sqrt{50}$ or $5\sqrt{2}$

   c. 4

   d. 4.664 . . .

   e. 3.24

   f. 4.032 . . .

6. **a.** 12

   b. 12

   c. 12

   d. 16

   e. 16

Discussion Answers

A. Answers will vary. The Pythagorean theorem will fail. In the case of acute triangles, the large square is less than the sum of the other two. In the case of obtuse triangles, it is greater.

B. In the case of horizontal distance, it is enough to take the absolute value of the difference between the x-coordinates. In the case of vertical distance, it is enough to take the absolute value of the difference between the y-coordinates.

C. The following is a list of the vertices other than (0, 0). More can be found by switching the x- and y-coordinates.

Legs of length 5:

(5, 0) (4,3)

(5, 0) (3, 4)

Legs of length 10: double the above coordinates.

Legs of length $\sqrt{50}$:

(5, 5) (7, 1)

Legs of length $\sqrt{65}$:

(8, 1) (7, 4)

(8, 1) (4, 7)

Legs of length $\sqrt{85}$:

(9, 2) (7, 6)

(9, 2) (6, 7)

Lab 9.3: Simplifying Radicals

Prerequisites: Once again, Lab 8.5 (Geoboard Squares) is required.

Simple radical form is no longer as important as it used to be as an aid to computation, since it is actually no simpler to enter $7\sqrt{3}$ into your calculator than $\sqrt{147}$. Conceptually, however, this technique remains important, as it is impossible to fully understand square roots and radicals without understanding this method. Moreover, simple radical form is often useful in problem solving and in communication, as it makes it easier to recognize certain common square roots. The geometric approach presented in this lab helps anchor the technique in a visual image. This does not make the actual manipulation easier to do, but it does help students to see it in a different way, it makes it easier to retain for some students, and it gives everyone a deeper understanding. Availability of technology is not an excuse for teaching less—it is an opportunity to teach more.

Answers

1. **a.** 10

   b. $\sqrt{10}$

   c. 40

   d. See Problem 2.

2. The side of the large square can be thought of as $\sqrt{40}$ or as twice the side of the small square: $2\sqrt{10}$.

3.

4.

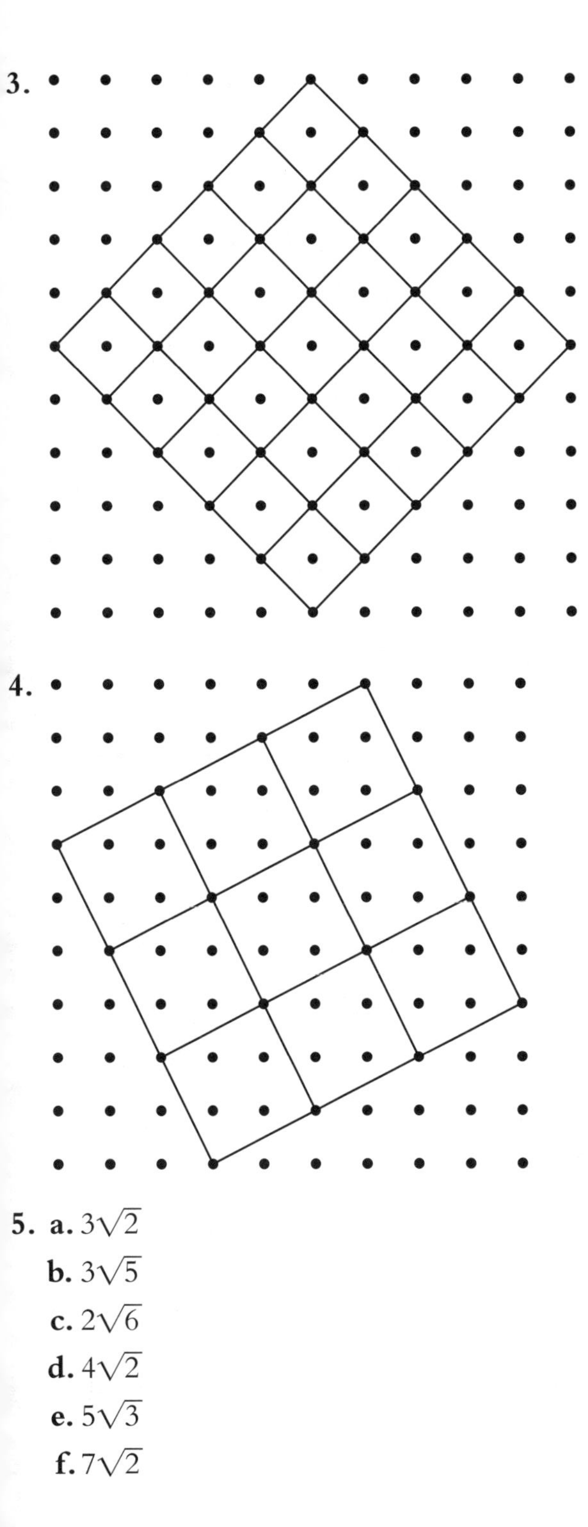

5. **a.** $3\sqrt{2}$

b. $3\sqrt{5}$

c. $2\sqrt{6}$

d. $4\sqrt{2}$

e. $5\sqrt{3}$

f. $7\sqrt{2}$

Discussion Answers

A.

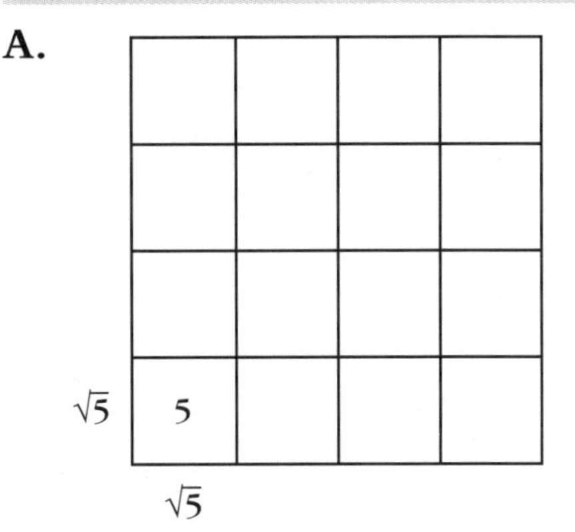

The side of the large square is $4\sqrt{5}$. Its area is $5 \cdot 16 = 80$. It follows that $4\sqrt{5} = \sqrt{80}$.

B. If the square of m is the greatest square that is a factor of n, we have $n = km^2$ for some number k. It follows that a square of area n could be divided into m^2 squares of area k. So $\sqrt{n} = m\sqrt{k}$. (Here, n, m, and k are positive whole numbers.)

Lab 9.4: Distance from the Origin

Prerequisites: Familiarity with the Pythagorean theorem, simple radical form, and slope. This lab makes connections among all these concepts and works well as a wrap-up report on the work done in Labs 8.5, 9.1, 9.3, and 9.4.

Students may use the records of their work from Lab 8.5 (Geoboard Squares), but don't suggest it if they don't think of it. Instead, encourage them simply to apply the Pythagorean theorem many times. This provides a useful, nonrandom drill. Working with neighbors and looking for patterns should speed up the process.

Answers

1.

10	10	$\sqrt{101}$	$2\sqrt{26}$	$\sqrt{109}$	$2\sqrt{29}$	$5\sqrt{5}$	$2\sqrt{34}$	$\sqrt{149}$	$2\sqrt{41}$	$\sqrt{181}$	$10\sqrt{2}$
9	9	$\sqrt{82}$	$\sqrt{85}$	$3\sqrt{10}$	$\sqrt{97}$	$\sqrt{106}$	$3\sqrt{13}$	$\sqrt{130}$	$\sqrt{145}$	$9\sqrt{2}$	$\sqrt{181}$
8	8	$\sqrt{65}$	$2\sqrt{17}$	$\sqrt{73}$	$4\sqrt{5}$	$\sqrt{89}$	10	$\sqrt{113}$	$8\sqrt{2}$	$\sqrt{145}$	$2\sqrt{41}$
7	7	$5\sqrt{2}$	$\sqrt{53}$	$\sqrt{58}$	$\sqrt{65}$	$\sqrt{74}$	$\sqrt{85}$	$7\sqrt{2}$	$\sqrt{113}$	$\sqrt{130}$	$\sqrt{149}$
6	6	$\sqrt{37}$	$2\sqrt{10}$	$3\sqrt{5}$	$2\sqrt{13}$	$\sqrt{61}$	$6\sqrt{2}$	$\sqrt{85}$	10	$3\sqrt{13}$	$2\sqrt{34}$
5	5	$\sqrt{26}$	$\sqrt{29}$	$\sqrt{34}$	$\sqrt{41}$	$5\sqrt{2}$	$\sqrt{61}$	$\sqrt{74}$	$\sqrt{89}$	$\sqrt{106}$	$5\sqrt{5}$
4	4	$\sqrt{17}$	$2\sqrt{5}$	5	$4\sqrt{2}$	$\sqrt{41}$	$2\sqrt{13}$	$\sqrt{65}$	$4\sqrt{5}$	$\sqrt{97}$	$2\sqrt{29}$
3	3	$\sqrt{10}$	$\sqrt{13}$	$3\sqrt{2}$	5	$\sqrt{34}$	$3\sqrt{5}$	$\sqrt{58}$	$\sqrt{73}$	$3\sqrt{10}$	$\sqrt{109}$
2	2	$\sqrt{5}$	$2\sqrt{2}$	$\sqrt{13}$	$2\sqrt{5}$	$\sqrt{29}$	$2\sqrt{10}$	$\sqrt{53}$	$2\sqrt{17}$	$\sqrt{85}$	$2\sqrt{26}$
1	1	$\sqrt{2}$	$\sqrt{5}$	$\sqrt{10}$	$\sqrt{17}$	$\sqrt{26}$	$\sqrt{37}$	$5\sqrt{2}$	$\sqrt{65}$	$\sqrt{82}$	$\sqrt{101}$
0	0	1	2	3	4	5	6	7	8	9	10
	0	**1**	**2**	**3**	**4**	**5**	**6**	**7**	**8**	**9**	**10**

Discussion Answer

A. Answers will vary.

a. The distances are symmetric with respect to the $y = x$ line.

b. Multiples of a given distance all lie on two lines with reciprocal slopes. For example, the multiples of $\sqrt{5}$ are on the lines with slope 1/2 and 2, respectively.

Lab 9.5: Area Problems and Puzzles

Prerequisites: In this lesson, students apply the formula for the area of a triangle (area = base · height/2) as well as for the Pythagorean theorem to some "famous right triangles": the isosceles right triangle, the half-equilateral triangle, and the less illustrious 1, 2, $\sqrt{5}$ triangle. We will return to all of these from a variety of viewpoints in the next three sections.

Timing: This is not really a lab. It's a collection of problems, puzzles, and projects that can be used flexibly as enrichment.

Problem 1 reviews inscribed and central angles. See Lab 1.8 (The Intercepted Arc).

Problem 4 is a giant hint toward solving Problem 5.

Problem 6 is challenging. There are several stages in solving it: First, students need to figure out the side of the required square ($\sqrt{5}$); then they need to figure out how to use what they know about this to help them make the cuts; finally, they can try to improve their solution by minimizing the number of cuts. This last phase is optional, and students should certainly be considered successful if they find any solution.

The X can be done in just two cuts (and four pieces), as shown by this figure.

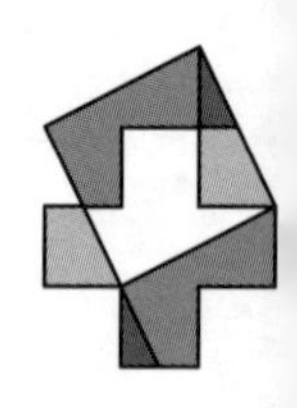

One way to find this solution is to superimpose two tilings of the plane: one tiling with the X, and one tiling with squares of area 5. The areas created by the overlapping figures form the dissection.

f your students enjoy Problem 6, similar puzzle is to dissect a entomino and rearrange the ieces to form two squares. The same approach used in Problem 6 works, as shown here.

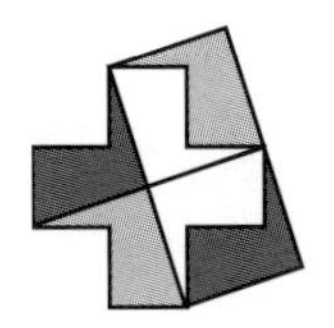

These puzzles are called dissections and are a art of ancient mathematics and recreational mathematics. A surprising result is that any olygon can be squared, that is, cut into pieces hat can be reassembled into a square. Your tudents may try their hands at doing this with ther polyominoes. A famous problem from ntiquity is the squaring of the circle, which was ater proved to be impossible.

Problem 7 is a substantial project. Students who ake it on should be encouraged to follow the trategy outlined in Lab 8.6 (Pick's Formula): eeping one variable constant and changing the ther to see how it affects the area.

Answers

1.

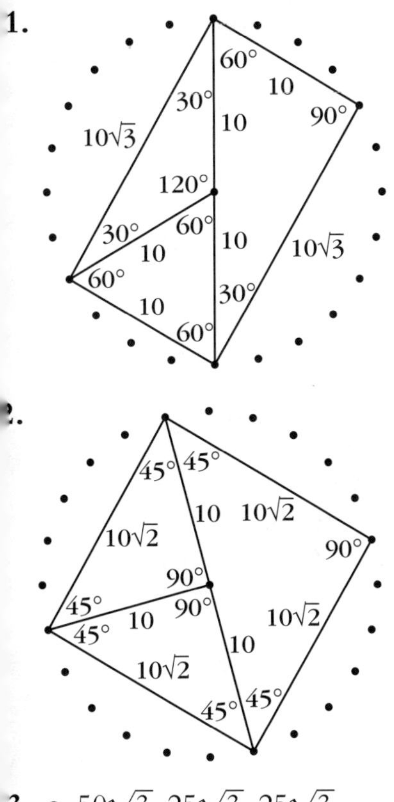

2.

3. **a.** $50\sqrt{3}, 25\sqrt{3}, 25\sqrt{3}$

b. 100, 50, 50

4. Two tan blocks, one green

Or: one orange, one green

5. Orange: 1

Green: $\frac{\sqrt{3}}{4}$

Yellow: $\frac{3\sqrt{3}}{2}$

Blue: $\frac{\sqrt{3}}{2}$

Red: $\frac{3\sqrt{3}}{4}$

Tan: $\frac{1}{2}$

6. See Notes to this lab.

7. Rectangular array: If the rectangles are $1 \times k$, the formula is $A = k\,(i + b/2 - 1)$.

Triangular array: $A = \frac{\sqrt{3}}{4}\,(2i + b - 2)$

There is no equivalent for Pick's theorem on a hexagonal lattice. You can see this in the counterexample given on the figure: It shows two triangles that have the same area and the same number of boundary dots but a different number of inside dots. If there were a version of Pick's theorem on a hexagonal lattice, it would require that the number of inside dots be the same in the two triangles.

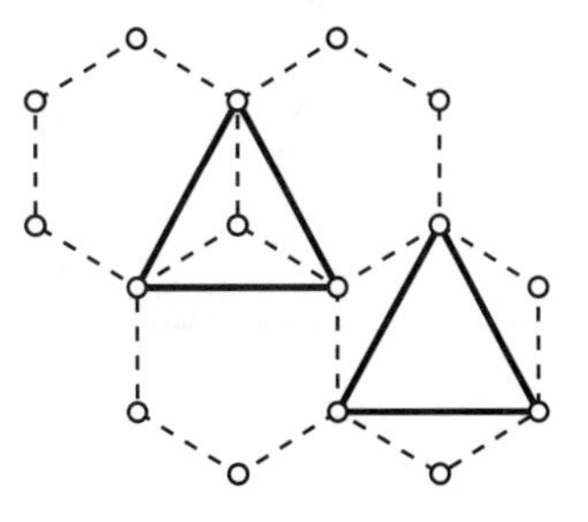

Lab 9.6: Taxicab Geometry

Prerequisites: Taxicab distance is introduced in Lab 9.1 (Taxicab Versus Euclidean Distance).

Timing: This is not really a lab, but a collection of problems that are suitable for in-class exploration or special student projects. To do it all would certainly take more than one period. If you want to do less than the full lesson, limit yourself to Problems 1–3 and Problems 6–8.

Geoboards are likely to be too small for this lesson.

Problem 1 should help students get into the taxicab world. A full discussion of it would have to include Question A.

Problem 2 is pivotal. Understanding the reason for the different cases is really fundamental to the remaining problems on the page.

The Euclidean concepts echoed in these problems are the *perpendicular bisector* (Problem 2), the *triangle inequality* (Problem 3 and Question B), the *ellipse* (Problem 4), and the *construction* of various circles (Problems 7–9). The Euclidean version of Problem 5 is more difficult. See "A New Look at Circles," by Dan Bennett, in *Mathematics Teacher* (February 1989).

For more problems in this domain, including taxi-parabolas, see *Taxicab Geometry,* by Eugene F. Krause (Dover, 1987).

Answers

1. Answers will vary. Samples include the following.

   a. (0, 0) (3, 1) (1, 3). Taxicab sides are 4, 4, 4. Euclidean sides are $\sqrt{10}, \sqrt{10}, 2\sqrt{2}$.

   b. (0, 0) (5, 1) (2, 4). Taxicab sides are 6, 6, 6. Euclidean sides are $\sqrt{26}, 3\sqrt{2}, 2\sqrt{5}$.

2. **a.**

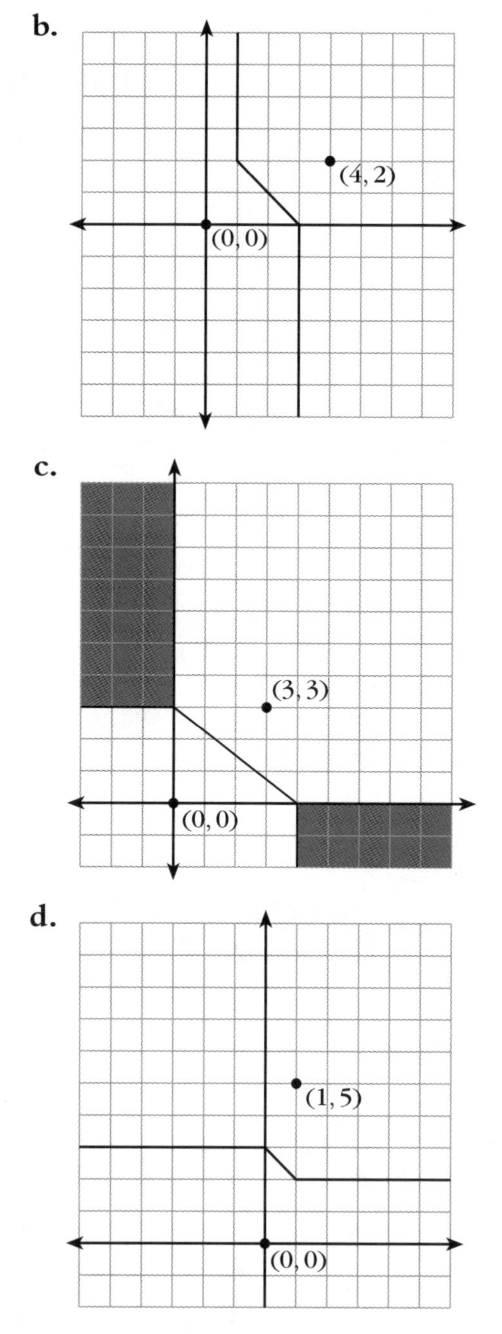

3. For Problem 3a, it is the line segment joining the points. For the other cases, it is the filled rectangle having the two points as opposite vertices.

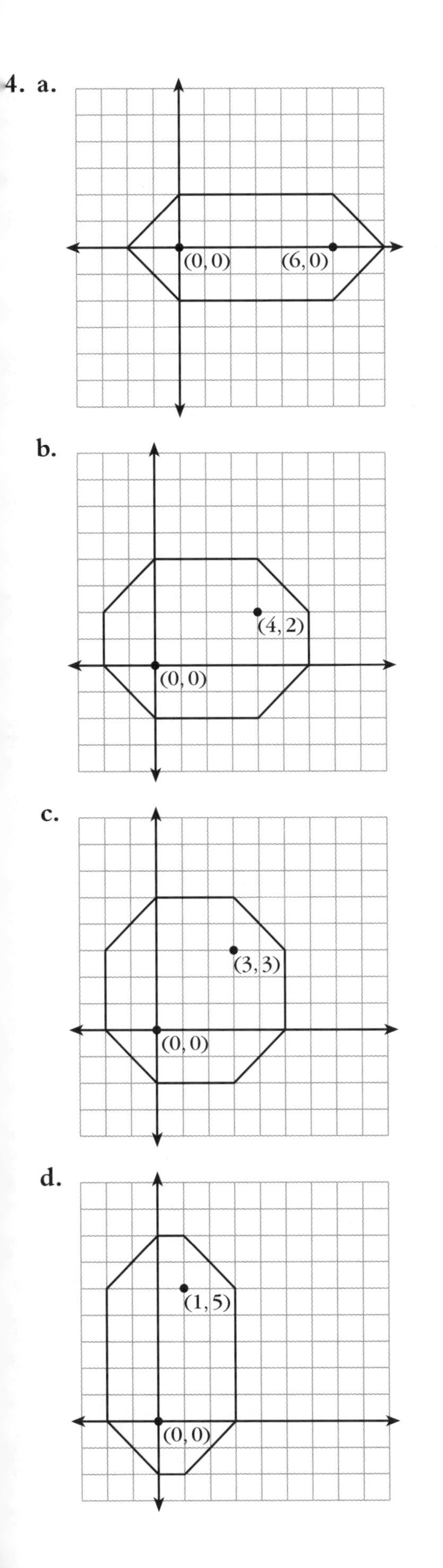
4. a.
(0, 0)
(6, 0)
b.
(4, 2)
(0, 0)
c.
(3, 3)
(0, 0)
d.
(1, 5)
(0, 0)

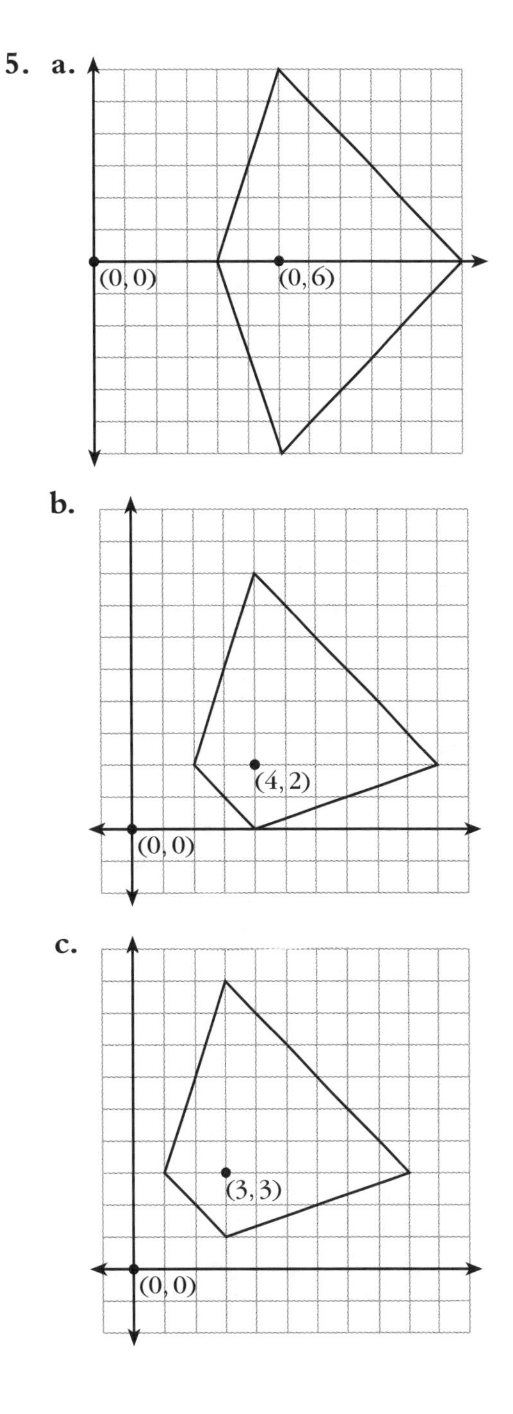
5. a.
(0, 0)
(0, 6)
b.
(4, 2)
(0, 0)
c.
(3, 3)
(0, 0)

d.

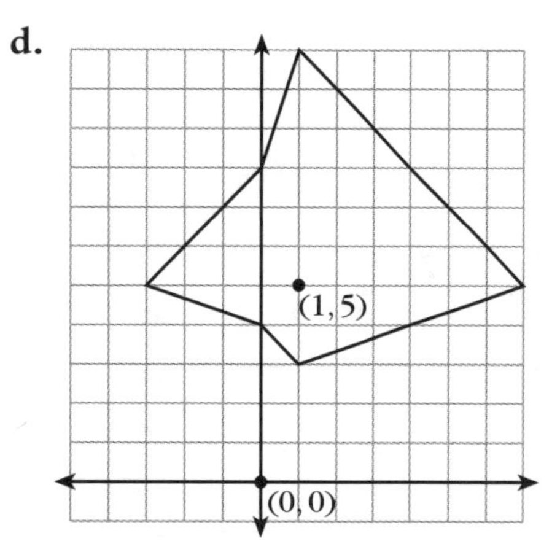

6. Taxi-π = 4
7. The centers are at (4, 3) and (5, 6).
8. The center can be at any point that is equidistant from A and B. (See Problem 2.) To actually draw the circle, find its radius first (the taxi-distance to A or B).
9. To explore this question, you need to explore different combinations of the cases from Problem 2. If the intersection of the sets of equidistant points is a single point, then that point is the center of the only suitable circle. However, in some cases the intersections can consist of no points (hence no center and no circle) or an infinite number of points (hence an infinite number of circles).

Discussion Answers

A. If $PA = PB$, then P can be any point on the sets you sketched in Problem 2. If $PA = AB$, then you can draw a taxi-circle centered at A with radius AB. Then P can be any point on that circle.

B. For two points, the set of points is the filled horizontal-vertical rectangle with the two points at opposite vertices.

10 Similarity and Scaling

Lab 10.1: Scaling on the Geoboard

Prerequisites: Students should be familiar with the concept of slope for Problems 4 and 5.

This lab introduces the concept of similarity and scaling and previews the theorems about the midpoints of triangles and quadrilaterals. It also provides a review of slope and distance.

The problem of constructing a segment whose midpoint is on a peg is interesting, and you may want to discuss it as a prelude (or postscript) to Problems 4 and 5, perhaps using the overhead geoboard to structure the conversation. One way to think about it is to note that the coordinates of the midpoint are the average of the coordinates of the endpoints. Therefore, if the midpoint is on a peg, the sum of the x-coordinates of the endpoints must be even, since it is twice the x-coordinate of the midpoint. But if the sum is even, then the difference must be even also; thus, the run is even. An analogous argument applies to the y-coordinates and the rise. The answer to Question C follows from these observations.

In Problem 5, students discover that the quadrilateral obtained by joining the midpoints of the sides an arbitrary quadrilateral is a parallelogram. While this fact, by itself, is not about similarity, it is included in this lab because its proof follows from the result discovered in Problem 4. See answers to Problems 4 and 5 and Question D.

Answers

1. To obtain (b) from (a), double the horizontal dimensions, or double the x-coordinates.

   To obtain (c) from (a), double the vertical dimensions, or double the y-coordinates.

   To obtain (d) from (a), double both the horizontal and vertical dimensions, or double both the x- and y-coordinates.
2. **a.** Copy (d) is scaled.

   b. Copy (b) is too wide; copy (c) is too tall.
3. The scaling factor for (a) to (d) is 2.

   The scaling factor for (d) to (a) is 1/2.
4. Answers will vary.

   a. The segment connecting two midpoints should be parallel to the other side and half as long. The small triangles should be congruent to each other and similar to the original triangle.

b. The scaling factor is 2 or 1/2. Other answers will vary.

5. Answers will vary.

a. The quadrilateral obtained by joining the midpoints of the sides of the original quadrilateral should be a parallelogram.

b. Answers will vary.

Discussion Answers

A. See answer to Problem 1.

B. They are reciprocals of each other.

C. It is impossible. One way to explain this is that if the midpoint of $\overline{AB}$ is on a peg, the rise and run from A to B are even, and likewise for $\overline{BC}$. The total rise and run for $\overline{AC}$ is obtained by adding the directed rises and runs of $\overline{AB}$ and $\overline{BC}$ and therefore must also be even. This forces the midpoint to be on a peg.

D. See answers to Problems 4 and 5. If you draw a diagonal of the quadrilateral, it is divided into two triangles. Using the result of Problem 4 on those triangles, you have two opposite sides of the inner quadrilateral that are both parallel to the diagonal and half as long. This is enough to guarantee a parallelogram.

Lab 10.2: Similar Rectangles

Prerequisites: The concept of slope is essential to Problems 3b and c and 5.

Timing: Problems 4 and 5 can be time-consuming, but they are well worth it.

This lab concentrates on similar rectangles. It has been my experience that (pedagogically) this is a more fundamental concept than the similar triangles we traditionally start with. Note that the lab provides an opportunity to review equivalent fractions, and their decimal representation, in a context that is new for many students.

At the same time, the lab reviews and reinforces the concept of slope, emphasizing the geometry of it rather than the concept of rate of change (which would, of course, be the core idea in an algebra, precalculus, or calculus course). Question D in particular provides an opportunity to discuss symmetry questions related to slope: Symmetry across the $y = x$ line corresponds to reciprocal slopes, and symmetry across the x- or y-axis corresponds to slopes opposite in sign. A detour is possible here to discuss the geometry of negative reciprocal slopes; they belong to perpendicular lines. This is a standard algebra 2 or precalculus topic that the geoboard makes accessible in earlier courses. This property is usually proved with the help of the Pythagorean theorem, but a perhaps more elegant way to get at the result is to observe that taking the negative of the reciprocal amounts to two reflections across lines that make a 45° angle to each other. This yields a 90° rotation.

The lab also helps prepare students for the next section, where the idea of the trigonometric tangent is founded on the concept of slope.

Answers

1. The angles are equal in both cases. The ratios of the sides are equal in (a): $2/3 = 4/6$, but they are not in (b): $2/3 \neq 5/6$.

2. The diagonal does not go through the vertex of the smaller rectangle.

3. a. (1, 2) (2, 4) (3, 6) (4, 8) (5, 10)

(2, 1) (4, 2) (6, 3) (8, 4) (10, 5)

b. 2

c. $\frac{1}{2}$

4. In addition to the set listed in 3a, we have the following:

(1, 1) (2, 2) (3, 3) (4, 4) (5, 5) (6, 6) (7, 7) (8, 8) (9, 9) (10, 10)

(1, 3) (2, 6) (3, 9) (3, 1) (6, 2) (9, 3)

(1, 4) (2, 8) (4, 1) (8, 2)

(1, 5) (2, 10) (5, 1) (10, 2)

(2, 3) (4, 6) (6, 9) (3, 2) (6, 4) (9, 6)

(2, 5) (4, 10) (5, 2) (10, 4)

(3, 4) (6, 8) (4, 3) (8, 6)

(3, 5) (6, 10) (5, 3) (10, 6)

(4, 5) (8, 10) (5, 4) (10, 8)

5. Geoboard slopes between 1 and 2, inclusive, are: 1; $1.\overline{1} = 10/9$; $1.125 = 9/8$; $1.\overline{142857} = 8/7$; $1.1\overline{6} = 7/6$; $1.2 = 6/5$; $1.25 = 5/4$; $1.\overline{285714} = 9/7$; $1.\overline{3} = 4/3$; $1.4 = 7/5$; $1.\overline{428571} = 10/7$; $1.5 = 3/2$; $1.6 = 8/5$; $1.\overline{6} = 5/3$; $1.75 = 7/4$; $1.8 = 9/5$; 2.

Discussion Answers

A. If rectangles are similar (and in the same orientation—vertical or horizontal), then the slopes of the diagonals should be the same. If the rectangles are nested so that they share a vertex, then the diagonals through that vertex should coincide.

B. If the point (a, b) is part of the answer, then so is (b, a).

C. It is easier to find the slopes as fractions ("rise over run"), but decimal notation makes it easier to compare the slopes with each other.

D. **a.** Use the reciprocals.

b. Use the opposites.

c. Use the reciprocals of the opposites.

Lab 10.3: Polyomino Blowups

Prerequisites: Students should know the definition of similar figures; Lab 10.1 (Scaling on the Geoboard) previews some of the ideas in this lesson.

Timing: Problems 7 and 8 could take a long time. They are engaging for many students, but they are not necessary to the lesson. A good compromise would be to use Problems 8b and 8c as extra-credit, out-of-class assignments.

Another approach to this lesson is to start with Problems 7 and/or 8, in order to create the motivation for Problems 1–6. Make sure students have a copy of the Polyomino Names Reference Sheet, from Section 4 (page 54), since it is unlikely that they will still remember the names of the various pieces.

Some students may prefer working all the problems on grid paper, without the cubes. That is quite all right.

For puzzle fiends among your students, here are some added challenges.

- The doubled tetrominoes can be tiled with only **I**'s.
- The doubled pentominoes can be tiled with mostly P's (only three N's are needed).
- The tripled pentominoes can be tiled with exactly five P's and four L's each.
- The tiling of quadrupled tetrominoes and pentominoes could be explored.

Not every tripled tetromino can be tiled with the same number of **l**'s and t's. This can be proved by a checkerboard argument. Color the tripled tetrominoes like a checkerboard. Note that the **l** will always cover an even number of black squares, and the t will always cover an odd number of black squares. In the tripled square, **l**, i, and n, there are eighteen black squares, so there's a need for an even number of t's in the final figure. However, in the tripled t, there is an odd number of black squares (seventeen or nineteen), so there's a need for an odd number of t's in the final figure.

Answers

1. Copy (d) is similar to the original because the measurements have been doubled in both dimensions, which means that the sides are proportional.

2. (In the Perimeter table, numbers in the Horiz. and Vertic. columns could be switched, depending on the orientation of the original polyomino.)

	Perimeter				Area			
		Doubled				Doubled		
	Original	Horiz.	Vertic.	Both	Original	Horiz.	Vertic.	Both
Monomino	4	6	6	8	1	2	2	4
Domino	6	10	8	12	2	4	4	8
Bent	8	12	12	16	3	6	6	12
Straight	8	14	10	16	3	6	6	12
Square	8	12	12	16	4	8	8	16
l	10	14	16	20	4	8	8	16
i	10	12	18	20	4	8	8	16
n	10	14	16	20	4	8	8	16
t	10	16	14	20	4	8	8	16

3. Answers will vary.
4. When measurements are doubled in both dimensions, the perimeter is doubled and the area is multiplied by four.
5. Predictions will vary.
6. Perimeter is multiplied by 3 or 4. Area is multiplied by 9 or 16.
7.

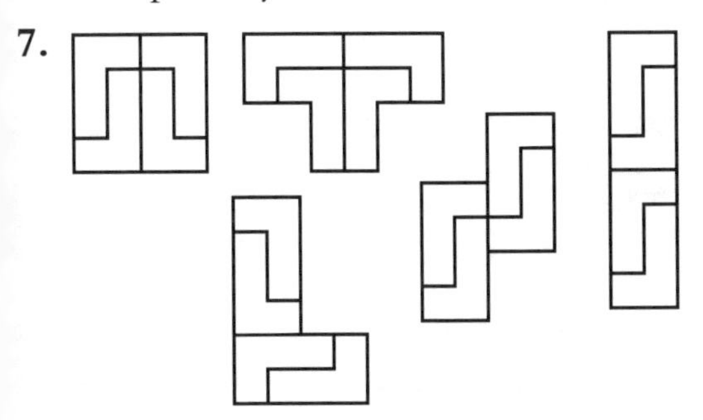

8. a.

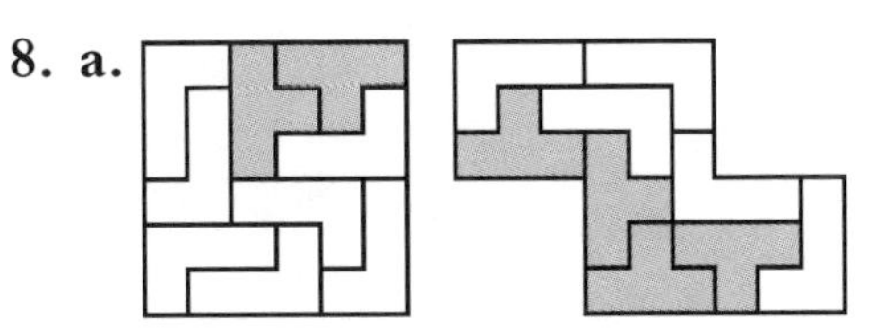

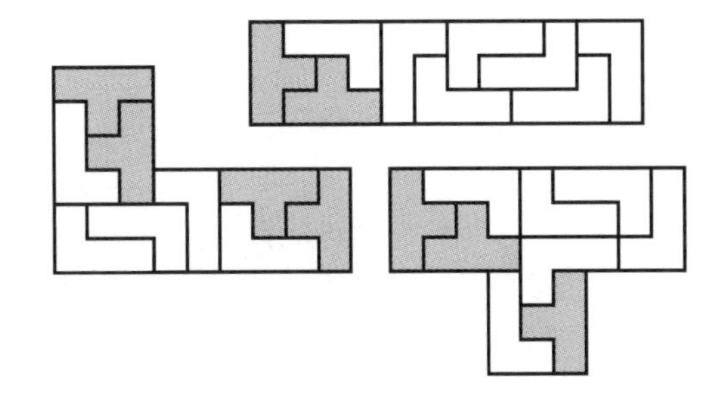

b.

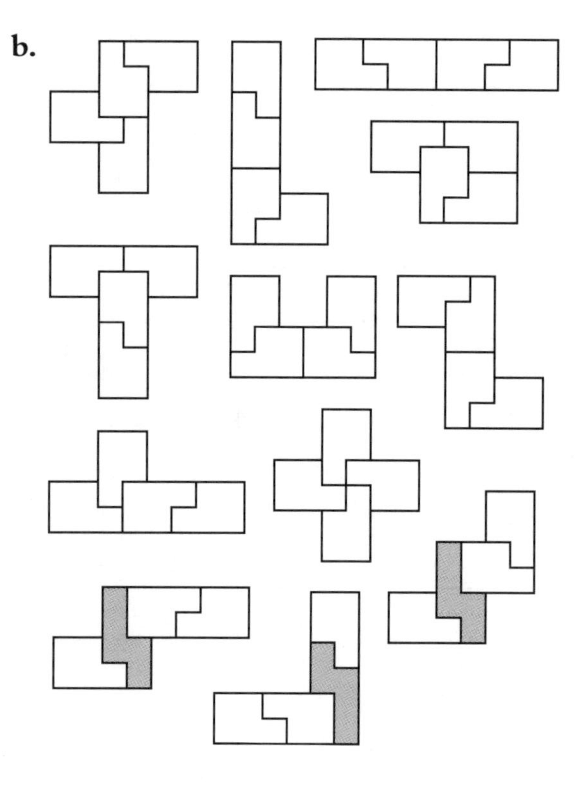

c.

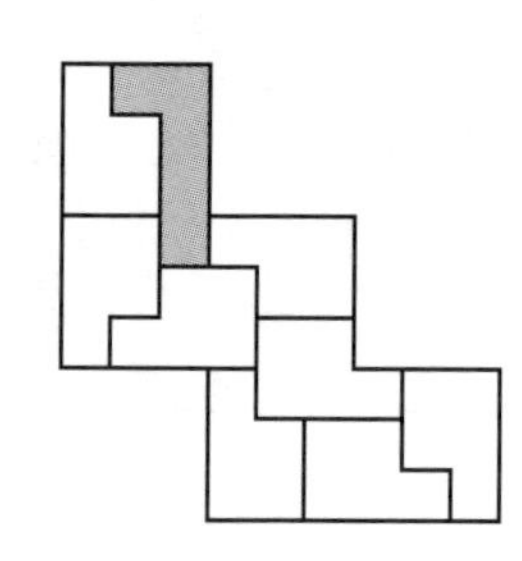

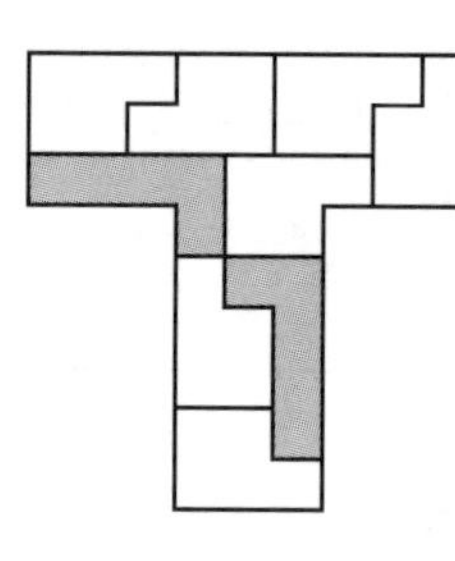

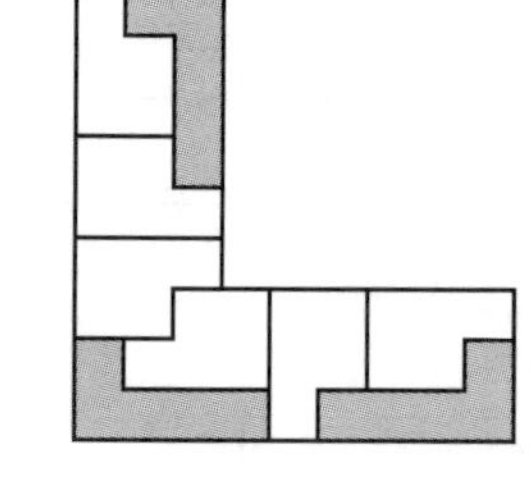

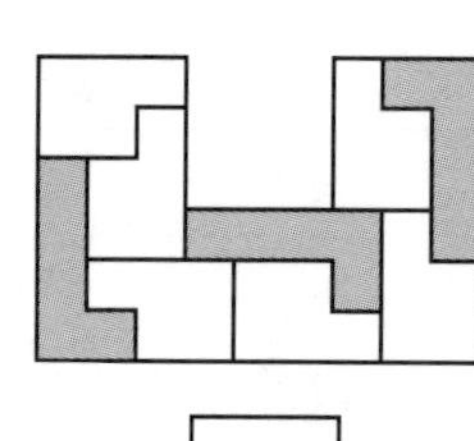

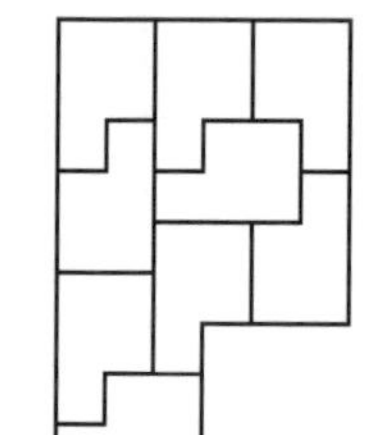

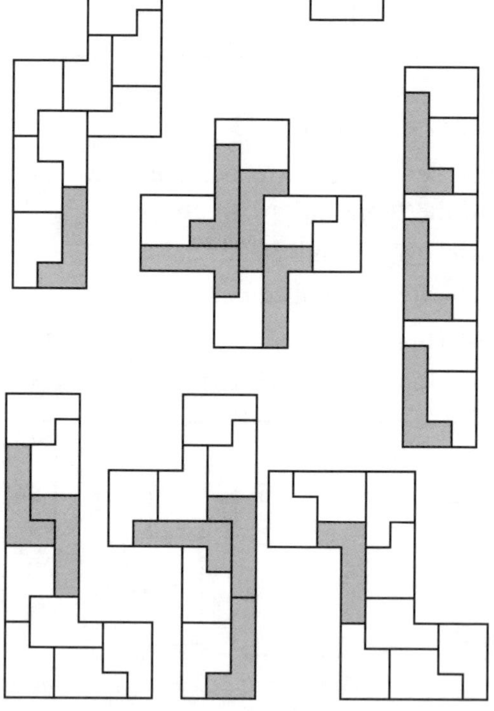

Discussion Answers

A. They are the same.

B. The ratio of areas is the square of the scaling factor. This is because the area is multiplied by the scaling factor twice, once for each dimension.

C. Nine. Because tripling a polyomino multiplies its area by $3^2 = 9$.

D. Tiling would require k^2 polyominoes, because multiplying the dimensions by k will multiply the area by k twice, once for each dimension.

Lab 10.4: Rep-Tiles

Prerequisites: Students should know the definition of similar figures; it would be helpful to have done some of the previous labs in this section.

This lab provides further practice with similarity and the ratio of areas in the context of now familiar shapes.

For Problems 1–3, students may prefer to work with pencil and eraser on grid paper rather than using the interlocking cubes.

Problem 3 is time-consuming and you may consider skipping it, since its conclusion is disappointing. It is challenging to make arguments about the impossibility of certain tilings. After the class has come to a conclusion about which polyominoes are rep-tiles, you may lead a discussion of why a given one, say the L, is not a rep-tile. The arguments start with forced moves: "I must place an L here this way, which forces me to place another one here, but now I see that if I place another L anywhere, the remaining space is not an L."

Problem 5 is based on the same figure as Problem 5 of Lab 10.1 (Scaling on the Geoboard), except that here we are working outward, while there we were working inward.

If students have trouble with Problem 6b, suggest that they cut out three copies of the triangle and work with those. If even then no one can solve the puzzle, discuss Question C, then draw the target triangle. Since the scaling factor is $\sqrt{3}$, use the hypotenuse as the short leg of the blown-up figure (because the hypotenuse is $\sqrt{3}$ times as long as the short leg in the original figure). The same sort of logic applies to Question D, which can be solved on grid or dot paper or on the geoboard.

Answers

1.

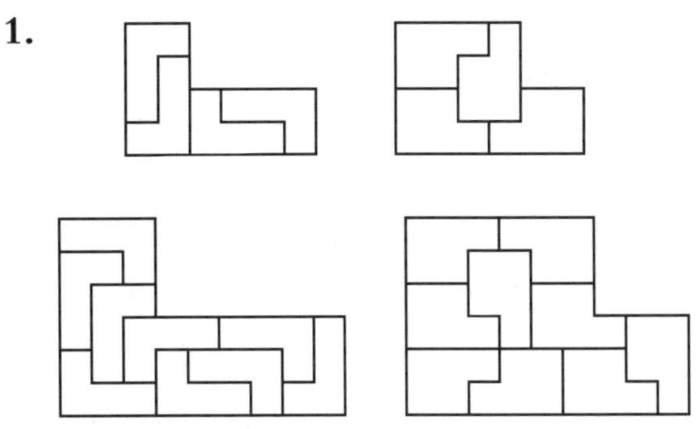

2. a. Four

b. Nine

3. Only the square and rectangular ones are rep-tiles.

4. Only the hexagon is not a rep-tile.

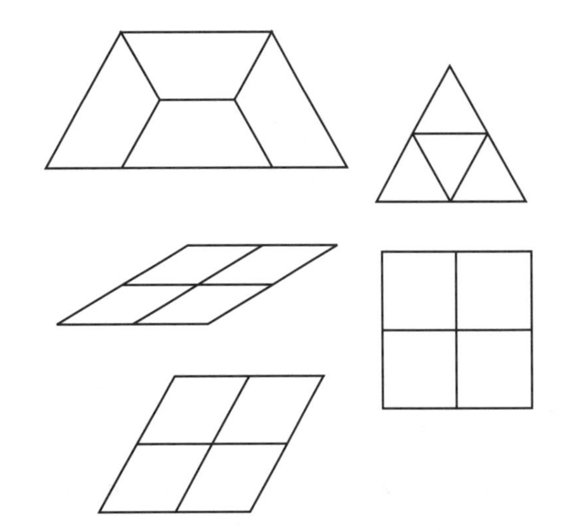

5. Surround it with three upside-down versions of itself.

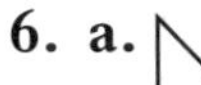

6. a.

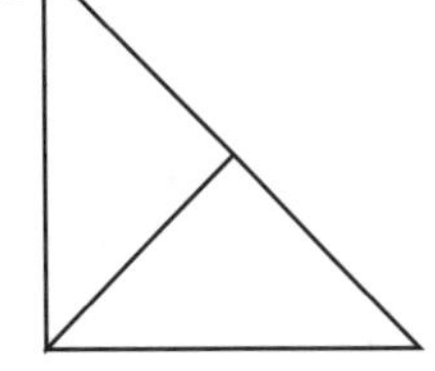

b.

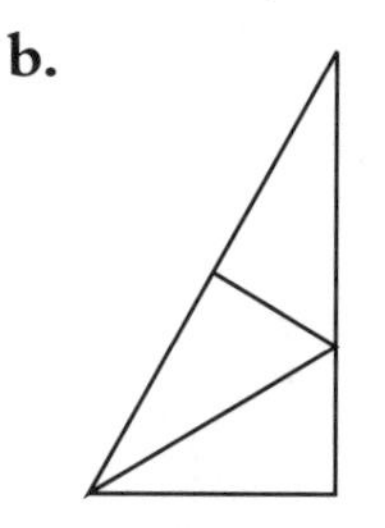

7. Only the parallelograms (including rhombi, rectangles, and squares) are rep-tiles.

Discussion Answers

A. The area is multiplied by k^2. See Lab 10.3 (Polyomino Blowups).

B. The method in Problem 5 will work for any triangle, not just the ones on the template.

C. The scaling factors are $\sqrt{2}$ and $\sqrt{3}$.

D. The scaling factor must be $\sqrt{5}$. This suggests that the 1, 2, $\sqrt{5}$ right triangle may work. Indeed, it does.

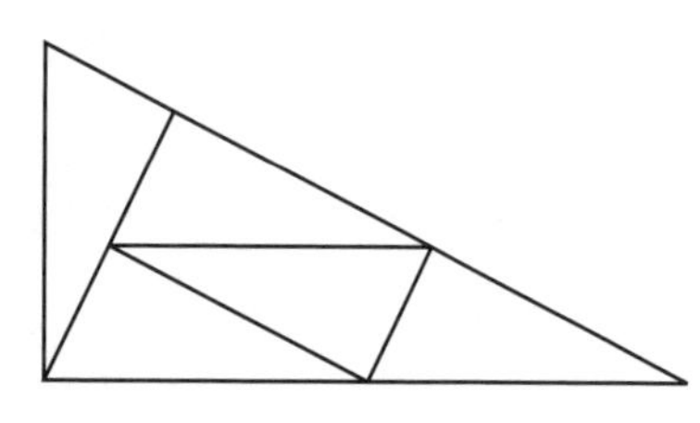

Lab 10.5: 3-D Blowups

Prerequisites: Students should know the definition of similar figures; Lab 10.3 (Polyomino Blowups) previews some of the ideas in this lesson in two dimensions.

Timing: Building 3-D blowups with the cubes can be quite time-consuming, especially if the original figure is large. Make sure students use a small number of blocks in the initial figure, no more than, say, four.

For Problems 2 and 3, many students find it very difficult to scale figures correctly in three dimensions. You will need to provide a lot of individual support, perhaps enlisting the help of talented students who finish their work quickly. The usual problem is a failure to conscientiously multiply *every* dimension by the scaling factor. This is not necessarily an issue of understanding—concentration and perceptual acuity are also involved.

Students are likely to solve Question C by studying the tables of values. Here is an algebraic approach:

$$\begin{aligned} V &= k^3 V_0 \\ &= k^2 k V_0 \\ &= \frac{A}{A_0} k V_0 \\ &= \frac{V_0}{A_0} k A \end{aligned}$$

So the constant c mentioned in Discussion Answer C is the ratio of the initial volume to the initial surface area.

Answers

1. **a.** The first and the last solids are similar.
   b. 2
   c. 14 and 56
   d. 4
   e. 3 and 24
   f. 8
2. Answers will vary.
3. Answers will vary.
4. Solutions for the surface area and volume columns will depend on the solid chosen.

Scaling factor	Surface area	Area ratio	Volume	Volume ratio
1		1		1
2		4		8
3		9		27
4		16		64
5		25		125
6		36		216

Discussion Answers

A. Each individual cube is scaled. Each of its faces goes from an area of 1 to an area of k^2, since it is multiplied by k twice only (the multiplication in the direction that is perpendicular to that face does not affect its area). It follows that the ratio of surface areas is k^2. Similarly, the volume of each individual cube gets multiplied by k^3, since it is multiplied by k three times. It follows that the overall ratio of volumes must be k^3.

B. $A = k^2 A_0$ and $V = k^3 V_0$

C. The formulas will be different for each original solid but will all be in the form $V = ckA$, where c is a constant determined by the shape and size of the original solid.

Lab 10.6: Tangram Similarity

Prerequisites: This lab includes some review of many ideas, including perimeter and area of similar figures and operations with radicals. The Pythagorean theorem, or at least its application to the particular case of an isosceles right triangle, is also relevant.

Another way to organize this lab is to start with Problems 5 and 6, which will motivate the methodical analysis in Problems 1–4.

It is the operations with radicals that makes this lab quite difficult for many students. Allow any sort of use of calculators to facilitate the work, but ask students to use simple radical form for their answers. That notation facilitates communication and reveals some of the relationships that are obscured by the decimal approximations. Note, however, that a familiarity with these particular decimal approximations can be very useful to students who later study trigonometry and calculus, which is why you should not discourage the use of calculators.

It is easiest to answer Question D by thinking of the small triangles as having area 1. Then the total area of all tangram pieces is 16, which can be broken up as 15 + 1, 14 + 2, 13 + 3, 12 + 4, 11 + 5, 10 + 6, 9 + 7, and 8 + 8. This would yield ratios of area of 15, 7, 13/3, 3, 11/5, 5/3, 9/7, and 1. The scaling factors would have to be the square roots of these numbers. Only in the last case can the scaling factor actually exist for tangram figures.

Answers

1. See Table 1.
2. See Table 2.

Table 1

Legs			Perimeters				
Sm Δ	**Md Δ**	**Lg Δ**	**Sm Δ**	**Md Δ**	**Lg Δ**	**Square**	**Parallelogram**
1	$\sqrt{2}$	2	$2 + \sqrt{2}$	$2 + 2\sqrt{2}$	$4 + 2\sqrt{2}$	4	$2 + 2\sqrt{2}$
$\frac{\sqrt{2}}{2}$	**1**	$\sqrt{2}$	$1 + \sqrt{2}$	$2 + \sqrt{2}$	$2 + 2\sqrt{2}$	$2\sqrt{2}$	$2 + \sqrt{2}$
$\frac{1}{2}$	$\frac{\sqrt{2}}{2}$	**1**	$1 + \frac{\sqrt{2}}{2}$	$1 + \sqrt{2}$	$2 + \sqrt{2}$	2	$1 + \sqrt{2}$

Table 2

Legs			Areas				
Sm Δ	**Md Δ**	**Lg Δ**	**Sm Δ**	**Md Δ**	**Lg Δ**	**Square**	**Parallelogram**
1	$\sqrt{2}$	2	$\frac{1}{2}$	1	2	1	1
$\frac{\sqrt{2}}{2}$	**1**	$\sqrt{2}$	$\frac{1}{4}$	$\frac{1}{2}$	1	$\frac{1}{2}$	$\frac{1}{2}$
$\frac{1}{2}$	$\frac{\sqrt{2}}{2}$	**1**	$\frac{1}{8}$	$\frac{1}{4}$	$\frac{1}{2}$	$\frac{1}{4}$	$\frac{1}{4}$

3. The scaling factors are as follows.

   Small to medium, medium to large: $\sqrt{2}$

   Small to large: 2

   Large to medium, medium to small:

$$\frac{1}{\sqrt{2}} = \frac{\sqrt{2}}{2}$$

   Large to small: $\frac{1}{2}$

4. The ratios of areas are as follows.

   Small to medium, medium to large: 2

   Small to large: 4

   Large to medium, medium to small: $\frac{1}{2}$

   Large to small: $\frac{1}{4}$

5. The answers are the same as for Problem 3.
6. Answers will vary.

Discussion Answers

A. The ratio of area is often easier to find, since it can often be done by seeing how many copies of the pieces making up the smaller figure can be used to cover the larger figure.

B. Answers will vary. Possibilities include the following:

- using the fact that the parallelogram and the square have the same area as the medium triangle
- multiplying an entire row by the scaling factor or the ratio of areas to get the other rows

C. 1

D. It is possible only if the scaling factor is 1.

Lab 10.7: Famous Right Triangles

This lesson pulls together much that has been learned in this and previous sections and also previews the type of thinking that underlies the next section on trigonometry.

It is difficult to overestimate the importance of these triangles. Just as students should be familiar with the multiplication tables, the perfect squares up to 15^2, the value of π up to four places, and so on, they should know these triangles by heart and be ready to use them when they come up. This is both a school survival strategy, as these numbers tend to appear often in texts and tests, and a concrete anchor to the all-important Pythagorean theorem. In the end, it is a matter of mathematical literacy.

Beware, however, of memorization that is not based on understanding! A colleague of mine told me that he asks his students this question: "A 30°, 60°, 90° triangle is a 3, 4, 5 triangle: always, sometimes, or never?" This is a great question that can reveal a serious lack of understanding. Try it with your students. I use this question in tests and quizzes every year.

Much of the work we have done in this book aims to lay a foundation of understanding to support these important results. The next section will lead students to a point where they will be able to find the angles in the 3, 4, 5 triangle without waiting for a future trigonometry course—and they will see that the angles are not 30°, 60°, and 90°.

Students will find that it is easier and less risky to solve Problems 7–12 by memorizing the sides of the reference triangles and scaling them than to solve them by using the Pythagorean theorem.

Answers

1.

	Leg_1	Leg_2	Hypotenuse
a.	1	1	$\sqrt{2}$
b.	1	2	$\sqrt{5}$
c.	1	$\sqrt{3}$	2
d.	3	4	5
e.	5	12	13

2. Triangle a: 1, 1, $\sqrt{2}$

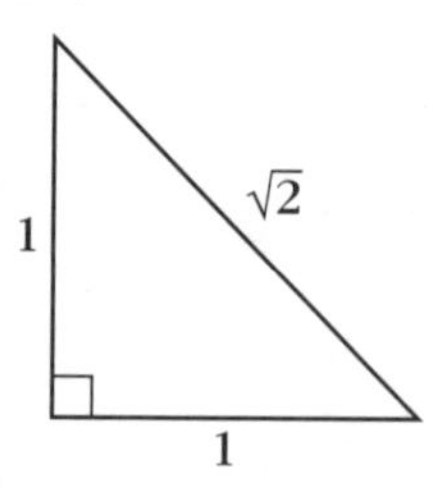

3. Triangle c: 1, $\sqrt{3}$, 2

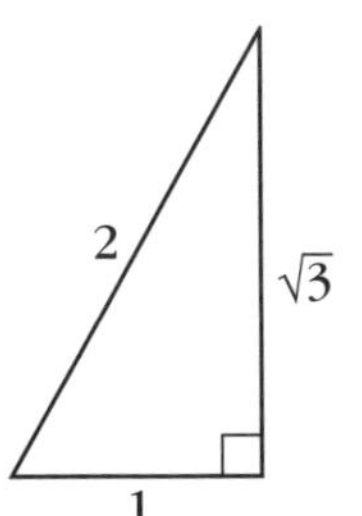

4. Triangles d and e

5.

	Leg_1	Leg_2	Hypotenuse
a.	2	2	$2\sqrt{2}$
b.	2	4	$2\sqrt{5}$
c.	2	$2\sqrt{3}$	4
d.	6	8	10
e.	10	24	26

6.

	Leg_1	Leg_2	Hypotenuse
a.	x	x	$x\sqrt{2}$
b.	x	$2x$	$x\sqrt{5}$
c.	x	$x\sqrt{3}$	$2x$
d.	$3x$	$4x$	$5x$
e.	$5x$	$12x$	$13x$

7. For triangle b: $10\sqrt{5}$; for triangle c: $10\sqrt{3}$

8. $60\sqrt{2}$

9. $\frac{30}{\sqrt{2}} = 15\sqrt{2}$

10. $35\sqrt{3}$

11. $\frac{80}{\sqrt{3}} = \frac{80\sqrt{3}}{3}$

12. 50 and $50\sqrt{2}$, or $25\sqrt{2}$ and $25\sqrt{2}$

Discussion Answers

A. Pattern blocks: the half-equilateral triangle (an understanding of which is necessary in order to find the area of most pattern blocks)

Tangrams: the isosceles right triangle

Geoboard: the 1, 2, $\sqrt{5}$ triangle; the 3, 4, 5 triangle

11 Angles and Ratios

Lab 11.1: Angles and Slopes

Prerequisites: Students must understand the concept of slope. Lab 10.1 (Scaling on the Geoboard) and Lab 10.2 (Similar Rectangles) are highly recommended.

To fill out the table, students may use rubber bands on the circle geoboard, as shown in the figures. Division should yield the slope, and the built-in protractor provides the angle. You should probably demonstrate the correct procedure on the overhead. When a rubber band simply connects two pegs, you can get an accurate reading off the protractor or ruler by looking at the tick mark that lies between the two sides of the rubber band. In other cases, such as when you form a rectangle or triangle with a rubber band, you will need to think about which side of the rubber band to look at for an accurate reading. If you choose correctly, you should get a reading accurate to the nearest degree or millimeter.

In Problem 4, students find angles when given slopes. Although students are directed to make slope triangles in the first and fourth quadrants and to give angles between –90° and 90°, some students are bound to make slope triangles in the second quadrant. In that case, they might (incorrectly) give angles measured from the negative x-axis, or they might (correctly) give angles between 90° and 180° that correspond to the given slopes. This is a good opportunity to point out that the angle corresponding to a given slope is not unique. In Problem 5, students find slopes when given angles. The slope corresponding to a given angle *is* unique. You can discuss how and why these situations are different as you discuss Question A.

Some students may ask if there is a way to find the angle or the slope with a calculator. If so, tell them that there is such a way, but that it is not the subject of that day's lab. In fact, the slope for a given angle is obtained with the "tan" key, and the angle between –90° and 90° for a given slope is found with the "$\tan^{-1}$" or "arctan" key. See the Trigonometry Reference Sheet and the section introduction for more on this.

If students are interested in getting more accuracy with the same apparatus, they can interpolate and extrapolate by using a string on the geoboard or a ruler on the paper copy of it (see page 245).

Answers

1–3. See student work.

4.

m	θ
0	0°
0.2	11°
0.4	22°
0.6	31°
0.8	39°
1	45°
1.25	51°
1.67	59°
2.5	68°
5	79°
None	90°
–5	–79°
–2.5	–68°
–1.67	–59°

5.

θ	m
0°	0.00
15°	0.27
30°	0.57
45°	1.00
60°	1.73
75°	3.73
90°	None
105°	–3.73
120°	–1.73
135°	–1.00
150°	–0.57
165°	–0.27
180°	0.00

Discussion Answers

A. Answers will vary.

The slopes of complementary angles are reciprocals of each other.

Slopes are positive for angles between 0° and 90° and between 180° and 270°.

The slope is 0 for 0° and 180°.

Slopes are negative for angles between 90° and 180° and between –90° and 0°.

Slopes are between 0 and 1 for angles between 0° and 45° and between 180° and 225°.

Slopes are greater than 1 for angles between 45° and 90° and between 225° and 270°.

B. The run is zero, and one cannot divide by zero.

C. Answers will vary. The first type of slope triangles work well for finding angles given slopes between 0 and 1 in the first table. The run stays a constant 10 cm (or 5 peg intervals), making it easy to identify the given slopes. It's necessary to use the second type of slope triangle to find angles given slopes whose absolute value is greater than 1. The third type of slope triangle will work for finding slopes given any angle in the second table. The fourth type is handy for finding slopes for given angles between 0° and 45° and between 135° and 180°. For those angles, once you read the rise off a vertical axis, finding the slope involves simply dividing by a run of 10.

D. The half-equilateral triangle is involved with both the 30° and 60° angles. The slopes obtained are consistent with the results of Lab 10.7, since $\sqrt{3}/1$ is close to 1.73 and $1/\sqrt{3}$ is close to 0.58.

The right isosceles triangle is involved with the 45° angle. The slope obtained is consistent with the results of Lab 10.7, since $1/1 = 1$.

Lab 11.2: Using Slope Angles

Prerequisites: Lab 11.1 (Angles and Slopes) is assumed.

This lab provides applications for the work in the previous lab. The discussion questions raise the issue of interpolating and extrapolating, since the tables we made in Lab 11.1 (Angles and Slopes) were clearly insufficient to deal with the full range of real-world problems of this type.

In doing this work, it is helpful to orient the figures so that the rise is vertical, the run is horizontal, and you are dealing only with positive slopes. Eventually, when the tangent ratio is officially introduced, the formulation "opposite over adjacent" will be more flexible, as it can be applied to a right triangle in any position.

After doing this lab, you can pursue any or all of the following three options if you want to do more work on this topic.

- Students can make a full table for slope angles, with 5° intervals. This is not as daunting as it seems; teamwork and patterns can help speed up the process. A method for doing it is described in Question A.
- You can use applications of the tangent ratio from any trig book. To solve the problems, students can use their circle geoboard or the circle geoboard sheet with a ruler to find the required ratios and angles, as suggested in the discussion.
- You can reveal that the slope ratio for a given angle is called the *tangent* of the angle and can be found on any scientific or graphing calculator, abbreviated as tan. The angle corresponding to a given slope is called the *arctangent* and is abbreviated as arctan or $\tan^{-1}$. A calculator set in degree mode will typically give an angle between –90° and 90°.

Answers

1. About 33 ft
2. About 30 m
3. **a.** See Problem 2.

   b. About 112 m
4. About 39°
5. About 31°
6. **a.** About 12.3 ft

   b. About 9.8 in.
7. About 233 m
8. About 67 units
9. See Problem 8.

Discussion Answers

A. Make a slope triangle. For example, in the case of the figure, a 35° angle yields a slope of 7/10. Conversely, a slope of 7/10 yields an angle of 35°.

B. 84°, 88°, 89°. As the slope increases, the angle gets closer to 90°.

Lab 11.3: Solving Right Triangles

Prerequisites: Lab 11.1 (Angles and Slopes) and Lab 11.2 (Using Slope Angles)

This lab raises a central problem of this whole section: solving right triangles. The work already done in the first two labs allows us to do just that, as it is possible to solve any right triangle by using only the tangent ratio and the Pythagorean theorem given one side and one other part (side or acute angle). However, Problem 7 shows that this is extraordinarily difficult in the case where no leg is known. This discovery motivates Lab 11.4, where we learn about the two ratios that involve the hypotenuse, thereby making Problem 7 completely accessible.

Note that the answers given are based on using only two significant digits, because of the limitations of using the circle geoboard to find tans and arctans. If you have already introduced these terms, much greater accuracy is possible with the help of calculators.

Answers

1. You can find the other acute angle (69°).
2. None
3. All other parts (hypotenuse: $\sqrt{41}$; angles: about 39° and 51°)
4. All other parts (other leg: $\sqrt{13}$, or about 3.6; angles: first find the slope ratio, 3.6/6 = 0.6, so the angles are about 31° and 59°)
5. All other parts (The other angle is 81°. The slope corresponding to the 9° angle is about 0.16, so the other leg is about 1.3 and the hypotenuse is about 8.1.)
6. All other parts (The other angle is 58°. The slope corresponding to that is about 1.6, so the other leg is about 16 and the hypotenuse is about 19.)
7. All other parts (The other angle is 25°. The slope for 65° is about 2.1. If the leg next to the 65° angle were 1, then the opposite leg would be 2.1 and the hypotenuse would be about 2.3. To find the actual legs, scale their assumed values by a factor of 4/2.3, which results in lengths of about 1.7 and 3.7.)

Discussion Answers

A. Two of the parts, including at least one side

B. If you know where the angles are, you know that the long leg is opposite the larger angle. If you only know the hypotenuse, you need to calculate the other leg. The legs would be equal if one angle is known to be 45° or if the leg and hypotenuse are in the ratio of 1 to $\sqrt{2}$.

C. The way we make a transition from angles to sides is with the help of the slope ratio, which involves the legs but not the hypotenuse. In Problem 7, we had no legs, just the hypotenuse.

Lab 11.4: Ratios Involving the Hypotenuse

Prerequisites: Lab 11.1 and Lab 11.2

This lab parallels Lab 11.1, but it is about the sine and cosine. Note, however, that we limit ourselves to angles between 0° and 90°. We will work with angles outside of that range in the discussion of Lab 11.6 (The Unit Circle).

Answers

θ	opp/hyp	adj/hyp
0°	0	1
15°	0.26	0.97
30°	0.50	0.87
45°	0.71	0.71
60°	0.87	0.50
75°	0.97	0.26
90°	1	0

opp/hyp	θ
0	0°
0.2	12°
0.4	24°
0.6	37°
0.8	53°
1	90°

adj/hyp	θ
0	90°
0.2	78°
0.4	66°
0.6	53°
0.8	37°
1	0°

Discussion Answers

A. Answers will vary. The opp/hyp and adj/hyp ratios are switched for complementary angles. When the angle is 0° or 90°, we no longer have a triangle, but we still have a ratio. At 0°, the opposite side is 0 and the adjacent side coincides with the hypotenuse, so the opp/hyp ratio is 0 and the adj/hyp ratio is 1. At 90°, this is reversed.

B. Right isosceles triangle: correct, since $1/\sqrt{2}$ is about 0.71

   Half-equilateral triangle: correct, since $1/2 = 0.5$ and $\sqrt{3}/2$ is about 0.87

   A less visible famous triangle here is the 3, 4, 5, which shows up with the ratios of 0.6 and 0.8. We now see that its angles are approximately 37° and 53°.

C. No. The hypotenuse is always longer than the legs, so the ratio cannot be greater than 1.

Lab 11.5: Using the Hypotenuse Ratios

This lab parallels Lab 11.2 (Using Slope Angles) but uses the sine and cosine ratios instead of the tangent ratio.

As mentioned before, most right triangle trigonometry problems cannot be solved with the tables we constructed. However, the methods discussed in Question A should make it possible to solve any problem of this type. For Question A, a piece of string or a ruler can be used to get horizontal or vertical lines between pegs, or you may prefer to use the paper version on page 245. The official introduction of trig terminology is on the Trigonometry Reference Sheet that follows this lab.

Answers

1. Height: about 1.3 cm. Area: about 4.6 cm^2.
2. About 4.8 cm^2
3. About 3.1 ft
4. About 37° and 53°
5. The height is 14.4/8, or 1.8 cm. Thus, opp/hyp = 1.8/3 = 0.6, so the angle is about 37°.
6. About 56 m

Discussion Answer

A. To find the opp/hyp ratio for 35°, draw a horizontal line at 35°, and observe that it meets the vertical axis at 5.7. It follows that the ratio is 5.7/10 = 0.57.

   For the adj/hyp ratio, use a vertical line.

   To find the angle when the ratio is known, start with the line and see where it meets the protractor.

Lab 11.6: The Unit Circle

Prerequisites: Students need the previous labs in this section and the Trigonometry Reference Sheet.

This is an introduction to the basic trig identities. After discussing them geometrically in this lab, it becomes possible to use algebraic notation such as $\sin(90 - x) = \cos x$ with some understanding. A good sequel to this lab is to make posters about all the identities to decorate the classroom. You may threaten to take the posters down at test time to impress upon the students the importance of knowing the identities. However, point out that it is far easier to reconstruct the unit circle sketches that accompany Questions D–G than it is to memorize the corresponding formulas. On the other hand, the Pythagorean identity and tan = sin/cos should be memorized and automatic.

Answers

1. For the tangent, calculate opp/adj in the larger triangle, where adj = 1.

   For the sine and cosine, calculate opp/hyp and adj/hyp in the smaller triangle, where hyp = 1.

2.

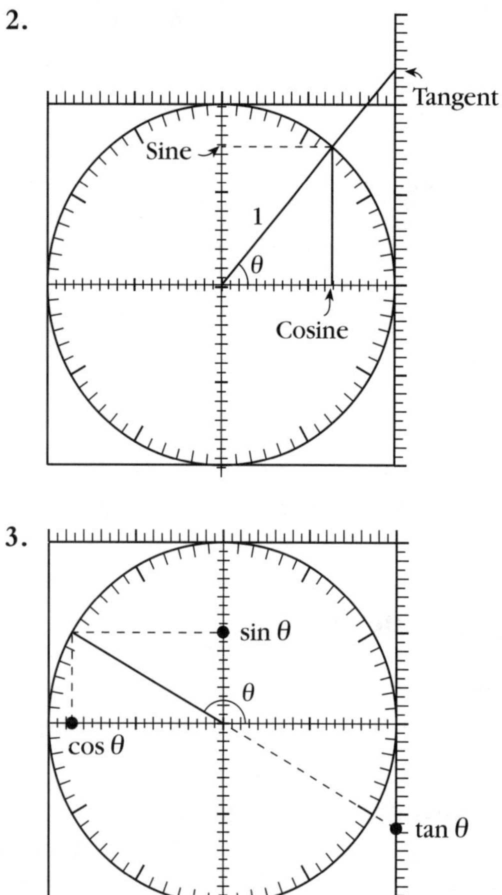

3.

4. The sines of the two angles are equal. Their cosines and tangents are opposite in sign.

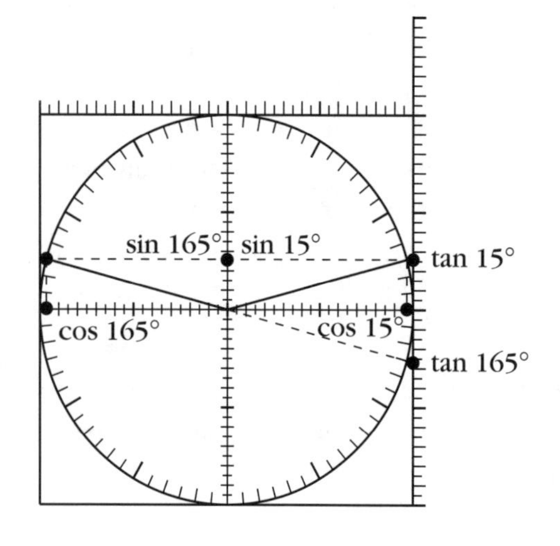

5. The cosines of the two angles are equal. Their sines and tangents are opposite in sign

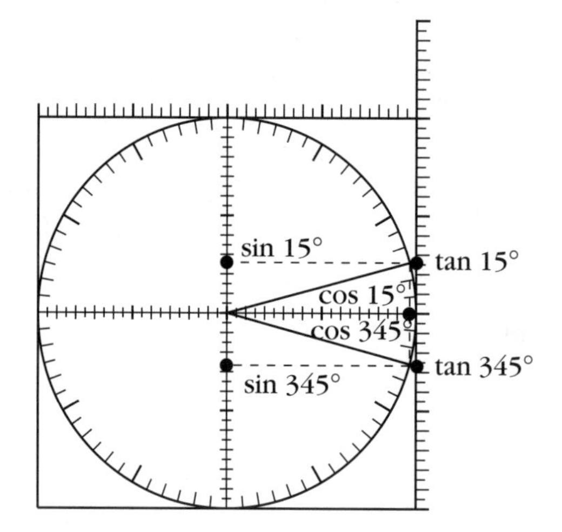

6. The tangents of the two angles are equal. Their sines and cosines are opposite in sign.

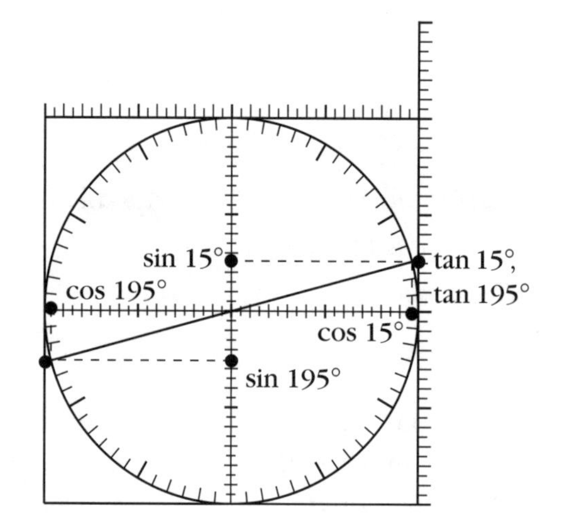

7. The tangents of the two angles are reciprocal The sine of one is the cosine of the other.

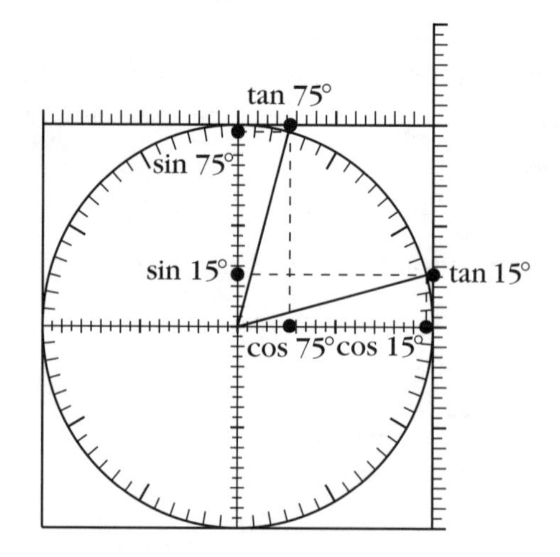

8. $\sin^2\theta + \cos^2\theta = 1$. It is the Pythagorean theorem in the smaller right triangle.

Discussion Answers

A. The ratios are the same because the two triangles are similar. See Problem 1 regarding which triangle is more convenient in each case.

B. $\tan = \frac{\text{opp}}{\text{adj}} = \frac{\sin}{\cos}$ in the small triangle

C. In the slope triangle shown below, angle θ intersects the top axis at a point $(c, 1)$. $\tan\theta = \text{opp}/\text{adj} = 1/c$. Therefore, $c = 1/\tan\theta$.

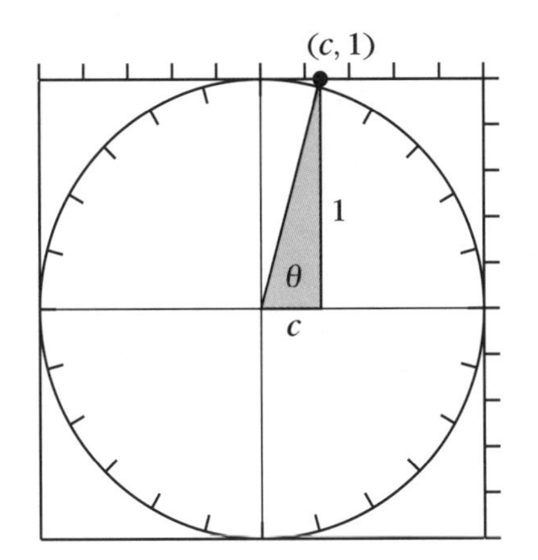

D. See Problem 4.

E. See Problem 7.

F. See Problem 6.

G. See Problem 5.

H. See Question G.

I. See Problem 7.

Lab 11.7: Perimeters and Areas on the CircleTrig Geoboard

Prerequisites: Right triangle trigonometry

Timing: Do not assign all of this in one sitting, as it could get tedious.

This lab applies basic trigonometry to various circle geoboard figures. It shows how much computing power is gained by having access to the trig ratios when measuring geometric figures. Most of the problems in this lab would be impossible to solve without trigonometry.

Students can approach it in different ways, depending on which angles they use for their calculations, but all approaches boil down to some version of the strategy suggested in Question A. Problems 1a and 1b, in that sense, are basic to this lab. This is the end of a long journey with the geoboard: The Pythagorean theorem made it possible to find any measurements on the Cartesian 11 × 11 geoboard, and now trigonometry makes it possible to find any measurements on the circle geoboard.

Note that Problem 3 and Questions A and B are good preparations for the next and final lab, where we will get more insight into the surprising whole number in Problem 3e.

Answers

1. a. $P = 38.48$; $A = 35.36$
   b. $P = 39.32$; $A = 25$
   c. $P = 51.67$; $A = 126.95$
2. a. $P = 44.35$; $A = 96.59$
   b. $P = 56.08$; $A = 193.18$
   c. $P = 56.08$; $A = 193.18$ (If you divide the rectangle in 2b along the horizontal diagonal and flip the bottom half, you get the figure in 2c—the perimeter and area are unchanged.)
3. a. $P = 51.96$; $A = 129.9$
   b. $P = 56.57$; $A = 200$
   c. $P = 60$; $A = 259.81$
   d. $P = 61.23$; $A = 282.84$
   e. $P = 62.12$; $A = 300$
   f. $P = 62.65$; $A = 310.58$

Discussion Answers

A. $P = 20 + 20\sin\theta$
$A = 100(\sin\theta)(\cos\theta)$

B. For the circle: $P = 62.83$ and $A = 314.16$, so the error in using the 24-gon figures is under 1 percent for the perimeter and under 2 percent for the area.

Lab 11.8: "π" for Regular Polygons

This final lab applies basic trigonometry to an interesting problem and makes a connection with a beautiful pattern block tessellation.
A historical note you can share with students: The classical method of computing π is to find the perimeters of inscribed and circumscribed polygons. Archimedes, ca. 240 B.C., used 96-gons to find that π is between 223/71 and 22/7. This gives π accurate to two decimal places.

The tessellation in Problem 3 (or at least its use to prove this result) was discovered by J. Kürschak in 1898 and brought to my attention by Don Chakerian one hundred years later. (See "Kürschak's Tile" by G. L. Alexanderson and Kenneth Seydel, in *The Mathematical Gazette,* 62 (1978), p. 192.)

If your students notice the pattern mentioned in the answer to Question A, they may be curious about why the pattern holds. This can lead to a discussion of the identity for $\sin 2\theta$, which is the underlying reason for this pattern.

Answers

1. a. $4r\sqrt{2}$

   b. $2\sqrt{2}$, or about 2.82

   c. $2r^2$

   d. 2

2. 3

3. The pieces covering the dodecagon can be rearranged to cover three of the dotted-line bounded squares (move three triangles and six half-tan pieces). But those squares have area r^2, so the area of the dodecagon is $3r^2$, and in this case $\pi_A = 3$.

4. π_P is about 2.94; π_A is about 2.38.

5.

n	π_P	π_A
3	2.5981	1.299
4	2.8284	2
5	2.9389	2.3776
6	3	2.5981
8	3.0615	2.8284
9	3.0782	2.8925
10	3.0902	2.9389
12	3.1058	3
16	3.1214	3.0615
18	3.1257	3.0782
20	3.1287	3.0902
24	3.1326	3.1058

Discussion Answers

A. Answers will vary.

One interesting observation is that π_P for n is equal to π_A for $2n$.

B. $\pi_P = n \sin(180°/n)$ and
$\pi_A = n \sin(180°/n) \cos(180°/n)$

C. The π's get closer to the actual value of π.

D. Since each value of n yields two values for polygon-π, there are too many values for any one of them to be important. On the other hand, the only way to find the perimeter and area of a circle is with the actual value of π.

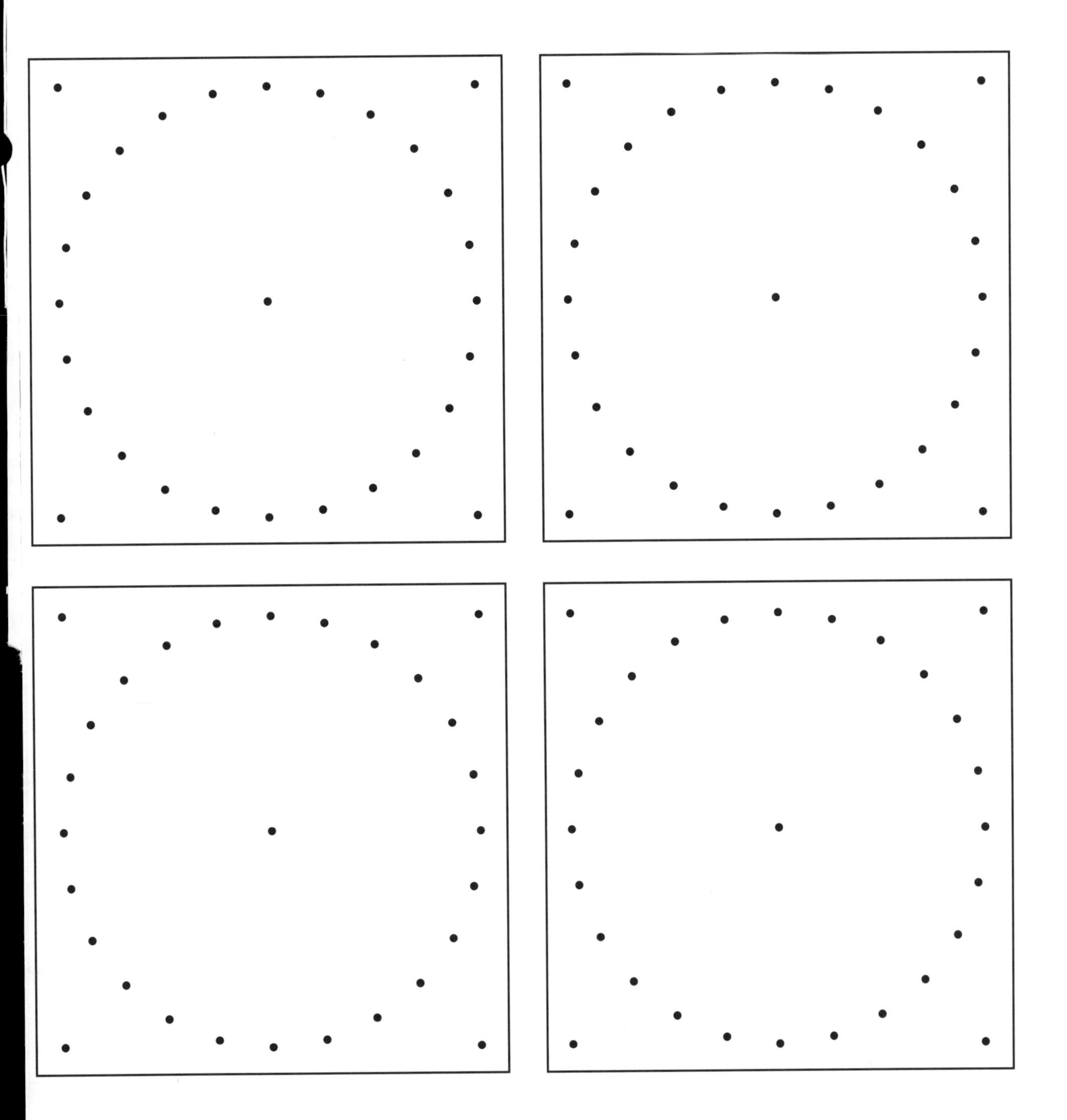

Instructions: Trace the figure onto a transparency. Cut along the solid lines (not the dashed lines). Tape the single square along an edge to form this shape:

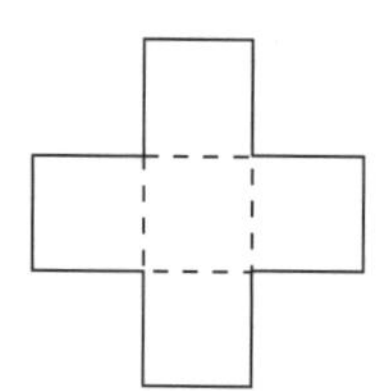

Fold to form an open cube. Tape remaining edges.

To make the radius of this circle exactly 10 cm, enlarge photocopies to 120%.

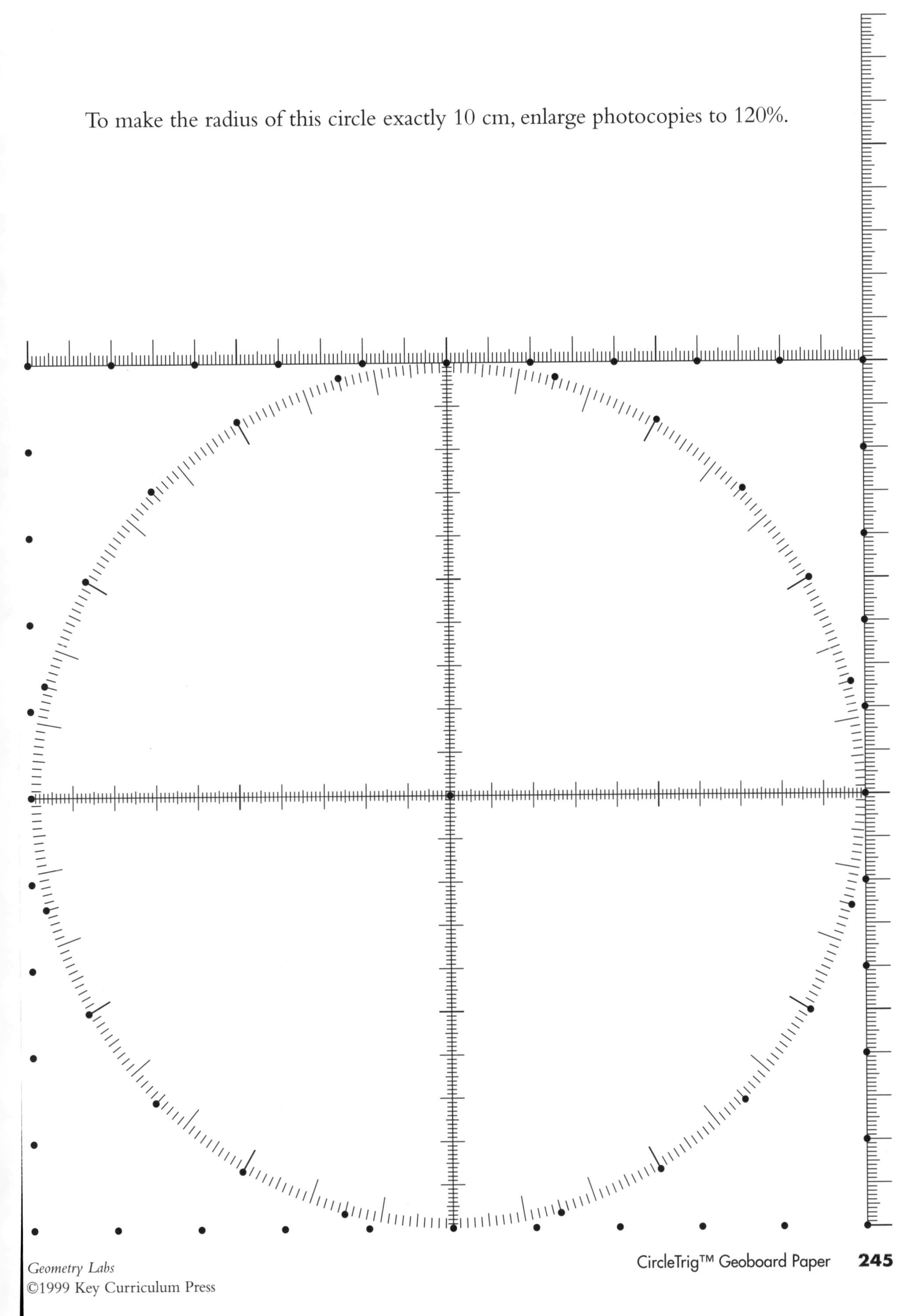

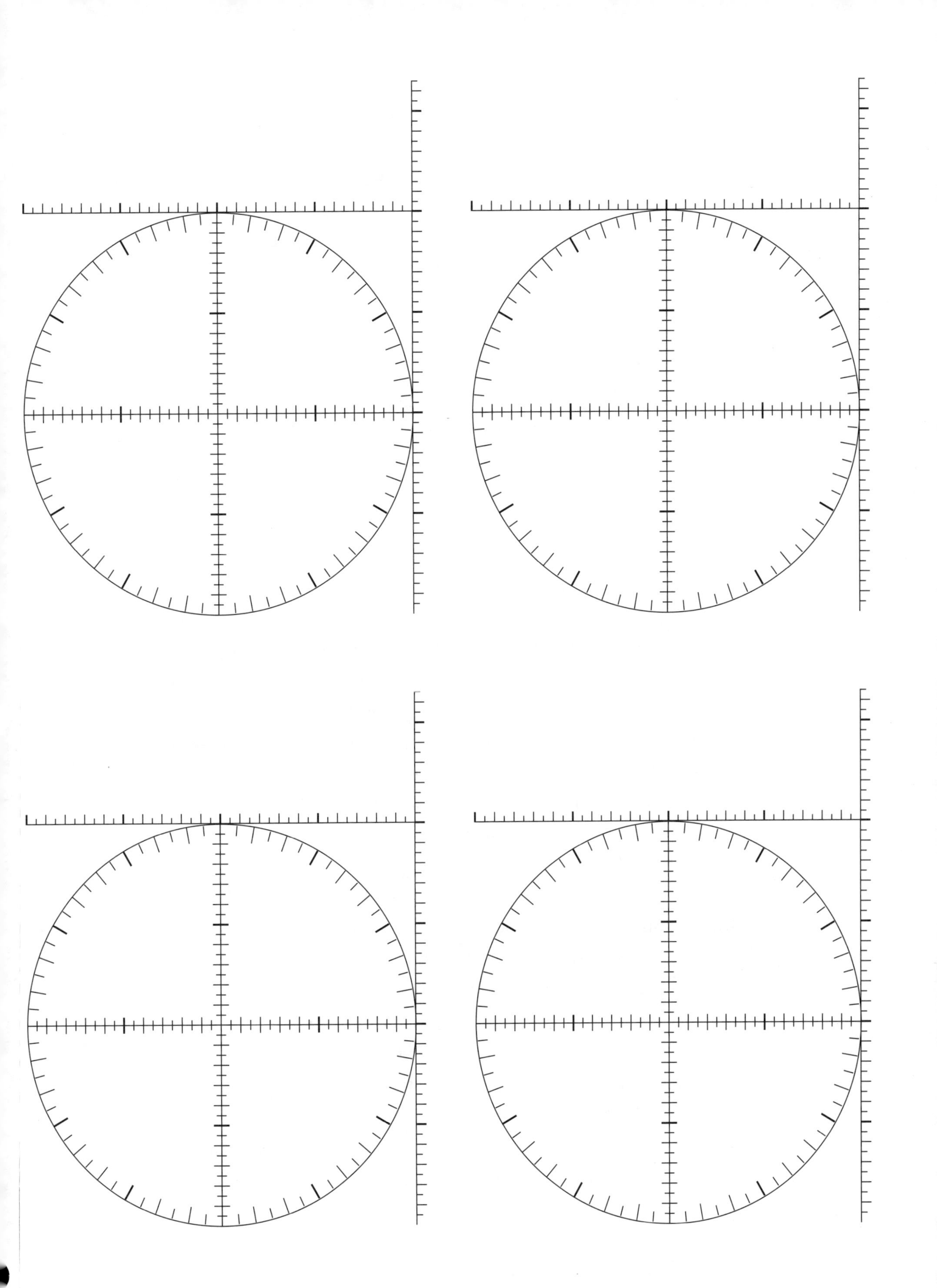

Bibliography

Alexanderson, G.L. and K. Seydel. "Kürschak's Tile." *The Mathematical Gazette,* vol. 62 (Oct. 1978) p. 192.

Bennett, D. "A New Look at Circles." *Mathematics Teacher,* vol. 82, no. 2 (Feb. 1989) pp. 90–93.

DeTemple, D.W. and D. A. Walker "Some Colorful Mathematics." *Mathematics Teacher,* vol. 89 (April 1996) p. 307ff.

Dudeney, H. *Amusements in Mathematics.* New York: Dover Books, 1958.

Gardner, M. *Mathematical Circus.* Chapters 5 and 15. Washington D. C.: Mathematical Association of America, 1992.

Gardner, M. "Mathematical Games." *Scientific American,* Recurring column.

Gardner, M. *Mathematical Magic Show,* Chapter 13. New York: Vintage Books, 1978.

Gardner, M. *The New Ambidextrous Universe.* New York: W. H. Freeman and Co., 1991.

Gardner, M. *New Mathematical Diversions from Scientific American.* Chapter 13. Washington D. C.: Mathematical Association of America, 1995.

Gardner, M. *Scientific American Book of Mathematical Puzzles and Diversions.* Chapter 13. Chicago: University of Chicago Press, 1987.

Gardener, M. *Sixth Book of Mathematical Diversions Scientific American.* Chapter 18. Chicago: University of Chicago Press, 1983.

Gardner, M. *Time Travel and Other Mathematical Bewilderments.* New York: W. H. Freeman and Co., 1987.

Golomb, S. W. *Polyominoes: Puzzles, Patterns, Problems, and Packings.* Princeton, N. J.: Princeton University Press, 1994.

Grunbaum, B. and G. C. Shephard. *Tilings and Patterns.* New York: W. H. Freeman and Co., 1987.

Krause, E. F. *Taxicab Geometry.* New York: Dover, 1987.

Picciotto, H. *Pentomino Activities, Lessons, and Puzzles.* Chicago: Creative Publications, 1984.

Picciotto, H. *SuperTangram Activities.* vol. I and II. Chicago: Creative Publications, 1986.

Serra, M. *Patty Paper Geometry.* Emeryville, Calif.: Key Curriculum Press.

Stevens, P. S. *Handbook of Regular Patterns: An Introduction to Symmetry in Two Dimensions.* Cambridge, Mass.: MIT Press, 1980.

Wah, H. and H. Picciotto. *Algebra: Themes, Tools, Concepts.* Chicago: Creative Publications, 1994.